INTERNATIONAL
WILDLIFE
ENCYCLOPEDIA

THIRD EDITION

Volume 15
POR–RIC

Marshall Cavendish Corporation
99 White Plains Road
Tarrytown, New York 10591–9001

Website: www.marshallcavendish.com

Library of Congress Cataloging-in-Publication Data

Burton, Maurice, 1898-
    International wildlife encyclopedia / [Maurice Burton, Robert Burton] .-- 3rd ed.
       p. cm.
    Includes bibliographical references (p.).
    Contents: v. 1. Aardvark - barnacle goose -- v. 2. Barn owl - brow-antlered deer -- v. 3. Brown bear - cheetah -- v. 4. Chickaree - crabs -- v. 5. Crab spider - ducks and geese -- v. 6. Dugong - flounder -- v. 7. Flowerpecker - golden mole -- v. 8. Golden oriole - hartebeest -- v. 9. Harvesting ant - jackal -- v. 10. Jackdaw - lemur -- v. 11. Leopard - marten -- v. 12. Martial eagle - needlefish -- v. 13. Newt - paradise fish -- v. 14. Paradoxical frog - poorwill -- v. 15. Porbeagle - rice rat -- v. 16. Rifleman - sea slug -- v. 17. Sea snake - sole -- v. 18. Solenodon - swan -- v. 19. Sweetfish - tree snake -- v. 20. Tree squirrel - water spider -- v. 21. Water vole - zorille -- v. 22. Index volume.
    ISBN 0-7614-7266-5 (set) -- ISBN 0-7614-7267-3 (v. 1) -- ISBN 0-7614-7268-1 (v. 2) -- ISBN 0-7614-7269-X (v. 3) -- ISBN 0-7614-7270-3 (v. 4) -- ISBN 0-7614-7271-1 (v. 5) -- ISBN 0-7614-7272-X (v. 6) -- ISBN 0-7614-7273-8 (v. 7) -- ISBN 0-7614-7274-6 (v. 8) -- ISBN 0-7614-7275-4 (v. 9) -- ISBN 0-7614-7276-2 (v. 10) -- ISBN 0-7614-7277-0 (v. 11) -- ISBN 0-7614-7278-9 (v. 12) -- ISBN 0-7614-7279-7 (v. 13) -- ISBN 0-7614-7280-0 (v. 14) -- ISBN 0-7614-7281-9 (v. 15) -- ISBN 0-7614-7282-7 (v. 16) -- ISBN 0-7614-7283-5 (v. 17) -- ISBN 0-7614-7284-3 (v. 18) -- ISBN 0-7614-7285-1 (v. 19) -- ISBN 0-7614-7286-X (v. 20) -- ISBN 0-7614-7287-8 (v. 21) -- ISBN 0-7614-7288-6 (v. 22)
    1. Zoology -- Dictionaries.   I. Burton, Robert, 1941- . II. Title.

    QL9 .B796 2002
    590'.3--dc21

                                                              2001017458

Printed in Malaysia
Bound in the United States of America

07 06 05 04 03 02 01   8 7 6 5 4 3 2 1

Brown Partworks
Project editor: Ben Hoare
Associate editors: Lesley Campbell-Wright, Rob Dimery, Robert Houston, Jane Lanigan, Sally McFall, Chris Marshall, Paul Thompson, Matthew D. S. Turner
Managing editor: Tim Cooke
Designer: Paul Griffin
Picture researchers: Brenda Clynch, Becky Cox
Illustrators: Ian Lycett, Catherine Ward
Indexer: Kay Ollerenshaw

Marshall Cavendish Corporation
Editorial director: Paul Bernabeo

Authors and Consultants

Dr. Roger Avery, BSc, PhD (University of Bristol)

Rob Cave, BA (University of Plymouth)

Fergus Collins, BA (University of Liverpool)

Dr. Julia J. Day, BSc (University of Bristol), PhD (University of London)

Tom Day, BA, MA (University of Cambridge), MSc (University of Southampton)

Bridget Giles, BA (University of London)

Leon Gray, BSc (University of London)

Tim Harris, BSc (University of Reading)

Richard Hoey, BSc, MPhil (University of Manchester), MSc (University of London)

Dr. Terry J. Holt, BSc, PhD (University of Liverpool)

Dr. Robert D. Houston, BA, MA (University of Oxford), PhD (University of Bristol)

Steve Hurley, BSc (University of London), MRes (University of York)

Tom Jackson, BSc (University of Bristol)

E. Vicky Jenkins, BSc (University of Edinburgh), MSc (University of Aberdeen)

Dr. Jamie McDonald, BSc (University of York), PhD (University of Birmingham)

Dr. Robbie A. McDonald, BSc (University of St. Andrews), PhD (University of Bristol)

Dr. James W. R. Martin, BSc (University of Leeds), PhD (University of Bristol)

Dr. Tabetha Newman, BSc, PhD (University of Bristol)

Dr. J. Pimenta, BSc (University of London), PhD (University of Bristol)

Dr. Kieren Pitts, BSc, MSc (University of Exeter), PhD (University of Bristol)

Dr. Stephen J. Rossiter, BSc (University of Sussex), PhD (University of Bristol)

Dr. Sugoto Roy, PhD (University of Bristol)

Dr. Adrian Seymour, BSc, PhD (University of Bristol)

Dr. Salma H. A. Shalla, BSc, MSc, PhD (Suez Canal University, Egypt)

Dr. S. Stefanni, PhD (University of Bristol)

Steve Swaby, BA (University of Exeter)

Matthew D. S. Turner, BA (University of Loughborough), FZSL (Fellow of the Zoological Society of London)

Alastair Ward, BSc (University of Glasgow), MRes (University of York)

Dr. Michael J. Weedon, BSc, MSc, PhD (University of Bristol)

Alwyne Wheeler, former Head of the Fish Section, Natural History Museum, London

Picture Credits

Contents

PORBEAGLE

A plump shark with a large crescent shaped tail, the porbeagle can survive in much colder waters than most other sharks.

THE PORBEAGLE SHARK lives in the Atlantic, as far north as Scandinavia and Iceland in the east and Newfoundland in the west. It ranges as far south as Argentina in the western Atlantic and to South Africa in the east. It is also found in the waters off South Georgia in the South Atlantic, around Australia and New Zealand and around Kerguelen Island in the southern Indian Ocean. The porbeagle is able to retain the heat generated by its muscles and recycle it to warm its body. In this way the shark can keep its body temperature as much as 20° F (11° C) warmer than the surrounding water. This enables the species to live in colder seas than those inhabited by other sharks.

The porbeagle occasionally grows up to 11½ feet (3.5 m) long, although a more common length is 6 feet (1.8 m). It is a plump and full-bodied fish that tapers sharply toward the tail fin, which has a crescent moon shape. The upper lobe of the tail is larger than the lower lobe. The back is dark blue gray, and the belly is white.

The snout is sharply pointed and overhangs a large crescent-shaped mouth in which the jaws are armed with three or four rows of slender awl-like teeth with small cusps at the bases. The lower teeth are directed upward or slightly backward. The gill slits are large, and the spiracles (breathing holes) are small. The pectoral fins and the first dorsal fin are large, but the second dorsal fin, the pelvic fins and the anal fins are small. A small white patch is visible on the rear edge of the first dorsal fin.

The porbeagle's unusual name is said to be a combination of "porpoise," from its general appearance, and "beagle," for the way it hunts.

Related sharks

There has been confusion in the past between the porbeagle and the shortfin mako, *Isurus oxyrinchus*. The mako is a larger shark, up to 13 feet (4 m) long, with a weight of up to 1,200 pounds (540 kg). It is more slender in the body and has a longer, more pointed snout. Its teeth have no

PORBEAGLE

CLASS	**Chondrichthyes**
ORDER	**Lamniformes**
FAMILY	**Lamnidae**
GENUS AND SPECIES	***Lamna nasus***

WEIGHT
Up to 465 lb. (210 kg)

LENGTH
**Up to 11½ ft. (3.5 m); usually about
6 ft. (1.8 m)**

DISTINCTIVE FEATURES
**Stout, plump body; pointed, conical snout;
slender, awl-shaped teeth; long gill slits;
crescent-shaped tail fin; blue gray above,
sometimes with dark spots and blotches;
whitish underside; small white patch on
rear of first dorsal fin**

DIET
Small shoaling fish, squid and other sharks

BREEDING
**Ovoviviparous (eggs hatch within female's
body). Number of young: 1 to 5.**

LIFE SPAN
Up to 30 years

HABITAT
**Over submerged banks and reefs; generally
near surface**

DISTRIBUTION
**Coastal waters in Atlantic; also in parts of
Indian and Pacific Oceans and Mediterranean**

STATUS
Vulnerable

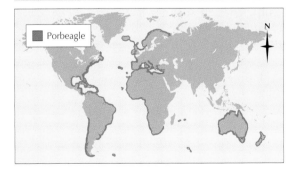

■ Porbeagle

Aggressive species

Both the porbeagle and the mako are aggressive
and dangerous to humans, the mako more so.
The mako has been known to attack small boats,
leaving some of its teeth behind in the timbers.
Indeed, the mako more than any other fish has
the reputation for doing this, yet there are no
records of its having attacked bathers in the
Atlantic and very few elsewhere, probably
because it does not come close enough to shore.

Swift hunters

The porbeagle is a voracious feeder, taking
mainly small fish. In the North Atlantic it feeds
on herring, cod, whiting, hake, mackerel and
dogfish. It also eats squid. The mako can over-
take a swordfish, which has been credited with
speeds of up to 60 miles per hour (96 km/h).
Like the porbeagle, the mako is a warm-blooded
shark, and scientists believe that such species are
capable of sustaining faster swimming speeds
than other shark species. When it has caught its
prey, the mako either bites off the fish's tail or
swallows the prey whole. There is one report of a
730-pound (330-kg) mako being caught with a
120-pound (54-kg) swordfish in its stomach.

The porbeagle is ovoviviparous, producing
eggs that hatch within the body of the mother.
There are usually one to five pups (young) in a
litter, measuring 28–32 inches (70–80 cm) long.

Porbeagles are not normally used for food,
and those that are caught are either thrown back
into the sea or used as fertilizer. However, they
are popular with sport fishers because of their
speed and strength in the water. The oil from
their livers is sometimes extracted, a 9-foot
(2.7-m) shark yielding up to 11 gallons (54 l) of
oil. At one time the Maori people of New
Zealand wore the teeth as personal ornaments,
often using them to adorn their ears.

*The shortfin mako
(above) is related
to the porbeagle.
Its streamlined body
minimizes water
resistance, enabling
it to conserve energy
when cruising. The
shark shown here
has been tagged by
scientists.*

small cusps. The shortfin mako ranges through
the whole Atlantic Ocean and is also found in the
Indian Ocean and South Pacific. The North
Pacific counterpart of the porbeagle is the salmon
shark, *Lamna ditropis*. Unlike the porbeagle, this
fish has dark blotches on its underside.

PORCUPINE FISH

To intimidate predators a porcupine fish draws in water and inflates itself like a balloon. At the same time sharp spines spring out all over its body. Pictured above is the long-spine porcupine fish.

WHEN IT IS RELAXED, a porcupine fish has a conventional fish shape. However, it is able to inflate itself to several times its normal body size until it becomes almost spherical, with long spines bristling all over it. The porcupine fish's tail and mouth appear very small compared with this greatly distended body. It has large eyes, and the dorsal, anal and pectoral fins are of moderate size. Porcupine fish grow to about 20 inches (50 cm) long. There are 19 species of porcupine fish, all belonging to the same family, Diodontidae, as the burrfish, which have short spines that are permanently erect.

Doubly armored

When they become disturbed or alarmed, porcupine fish inflate their bodies by drawing in water. The body swells, and the spines, which lie flat when the fish is calm, are erected, standing out almost at right angles to the surface of the skin. If a porcupine fish is suddenly taken out of water, it blows itself up by drawing in air. The spines, which may be up to 2 inches (5 cm) long, are sharp. Each of these long, stout spines has a three-armed base, the paired arms of which overlap in the skin with those of their fellows, providing a more or less continuous "coat of mail." In some species the spines have only two arms at the base and can be raised and lowered without the fish inflating itself.

Porcupine fish live in subtropical and tropical seas and are rather slow swimmers. They move by waving the dorsal and anal fins, helped to a small extent by the pectoral fins, and use the tail fin for steering.

The teeth of porcupine fish form a continuous plate in the upper jaw, with another plate in the lower jaw. Each plate has a sharp edge with a crushing surface behind it. The fish feed

LONG-SPINE PORCUPINE FISH

CLASS	**Osteichthyes**
ORDER	**Tetraodontiformes**
FAMILY	**Diodontidae**
GENUS AND SPECIES	***Diodon holacanthus***

ALTERNATIVE NAME
Spiny puffer

LENGTH
Up to 20 in. (50 cm)

DISTINCTIVE FEATURES
Large eyes; teeth form continuous plates in upper and lower jaws; moderately sized dorsal, anal and pectoral fins. When calm: typical, rather compact form. When disturbed or alarmed: body inflates into sphere; sharp spines up to 2 in. (5 cm) long project from skin.

DIET
Coral, mollusks, sea urchins and crabs

BREEDING
Hatching period: usually about 4 days

LIFE SPAN
Not known

HABITAT
Mainly over shallow coral reefs and rocks; occasionally over open, soft seabeds

DISTRIBUTION
Warm coastal waters in eastern Pacific, Atlantic and Caribbean

STATUS
Not threatened

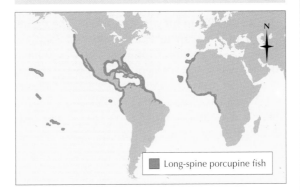

Long-spine porcupine fish

by crushing mollusks and by biting off and crushing pieces of coral. The flesh of the coral is digested in the stomach. The stony matter in the coral, crushed by the dental plates, accumulates over time; one dissected specimen had over 1 pound (0.5 kg) of crushed coral in its stomach.

Floating larvae

Prior to reproduction, the male porcupine fish slowly pushes the female to the water's surface. Spawning takes place at once, the eggs being spherical and buoyant, and hatching occurs after about 4 days. The larvae are mainly yellow in color, with scattered red spots, and well developed, with a functional mouth, eyes and gas bladder. Larvae that are less than 10 days old are covered with a thin shell; thereafter the shell is lost and they begin to develop spines.

The larvae undergo a metamorphosis about 3 weeks after hatching, when the teeth, fins and fin rays are formed. As they develop into juvenile fish, the larvae take on the characteristic coloration of adults, and dark spots appear on their underside. These spots may serve as camouflage to protect the juveniles from predators, such as dolphins, that swim below the seaweed. The spots remain until the juveniles move inshore and become adults. All remaining changes to the fish are external, and include elongation of the spines and body growth.

Mysterious behavior

There is still much scientific uncertainty as to the breeding habits or the predators of porcupine fish, and there are very few records of what happens when they are attacked. William Beebe, a distinguished American marine zoologist, observed some porcupine fish threatened by a 4-foot (1.2-m) garfish bunch together for protection. However, for no obvious reason, from time to time one left the mass and swam away, to be seized promptly and eaten by the garfish.

The combination of good camouflage and a sophisticated defensive strategy means that porcupine fish are likely to be safe from most marine predators.

PORPOISE

A Dall's porpoise, breaching. Porpoises are actually quite different from dolphins in a number of respects, including their small size, blunt snouts, spade-shaped teeth and triangular dorsal fin.

SOMETIMES THE NAMES porpoise and dolphin are used almost interchangeably to refer to any small whale. The situation is further complicated because the animal known as the common porpoise in the United States is called the bottlenose dolphin by the British, whereas the British common porpoise is known to Americans as the harbor porpoise. Often little more is seen than a tiny dorsal fin and a rolling back some distance out at sea, but with a good view it is quite easy to distinguish a porpoise from a dolphin. Most dolphins are larger than porpoises, which reach a maximum of 7¼ feet (2.2 m), but the main distinguishing features lie in the shape of the body. Porpoises have less streamlined bodies and blunt snouts, lacking the beaks of dolphins. The flippers are broad and rounded, whereas those of a dolphin tend to be curved and tapering. The back fin, the part most often seen, is small and triangular in the porpoise but more conspicuous, curving backward to a point, in the dolphin. In addition, porpoises have very different teeth to those of dolphins: porpoises have spade-shaped teeth compared to the conical teeth of dolphins.

Puffing pig and others

There are six species of porpoises. The harbor porpoise, *Phocoena phocoena*, is one of the most common species and is also known as the puffing pig. It has a dark gray back with paler gray patches on the flanks, and a white belly. The harbor porpoise lives in the eastern Atlantic, from the Arctic Sea to West Africa; in the western Mediterranean; in the western Atlantic, from Greenland to North Carolina; and in the eastern Pacific, from Alaska to California. The vaquita, *P. sinus*, is almost identical in appearance to the harbor porpoise but slightly darker. It lives in the Gulf of California, Mexico. Burmeister's porpoise, *P. spinipinnis*, is found on both sides of South America, from the Plate River, between Uruguay and Argentina, to Peru. The spectacled porpoise, *Australophocaena dioptrica*, ranges from the Plate River north to South Georgia. It is black above and white below, with a black rim around the eyes. Finally there is the finless porpoise, *Neophocaena phocaenoides*, found along the coasts and rivers of India and China, and Dall's, or True's, porpoise, *Phocoenoides dalli*, which lives in the North Pacific.

Found along coasts

Porpoises live in coastal waters and bays, and are often found in estuaries, sometimes penetrating far up rivers. One porpoise was found as far up the Rhine as Cologne. Porpoises live in pairs or in schools of up to 100 animals and are migratory. Little is known about the courses and timing of these migrations, but they seem to be most numerous in the North Sea in July and

PORPOISES

CLASS	**Mammalia**
ORDER	**Cetacea**
FAMILY	**Phocoenidae**

GENUS AND SPECIES **Harbor porpoise,** *Phocoena phocoena*; **Burmeister's porpoise,** *P. spinipinnis*; **vaquita,** *P. sinus*; **spectacled porpoise,** *Australophocaena dioptrica*; **Dall's porpoise,** *Phocoenoides dalli*; **finless porpoise,** *Neophocaena phocaenoides*

ALTERNATIVE NAMES
Harbor porpoise: common porpoise; puffing pig; Dall's porpoise: True's porpoise

WEIGHT
66–275 lb. (30–125 kg)

LENGTH
4–7¼ ft. (1.2–2.2 m)

DISTINCTIVE FEATURES
Blunt snout with indistinct or absent beak; spade-shaped teeth; triangular dorsal fin

DIET
Mainly fish; also crustaceans and squid

BREEDING
***Phocoena* species. Age at first breeding: 3–4 years; breeding season: mainly June–October; number of young: 1; gestation period: about 330 days; breeding interval: 1–2 years.**

LIFE SPAN
Usually up to 10 years

HABITAT
Offshore coastal waters; large rivers and bays

DISTRIBUTION
Harbor porpoise: North Atlantic and North Pacific; also western Mediterranean

STATUS
Common: harbor porpoise; endangered: finless and Burmeister's porpoises; critically endangered: vaquita; not known: Dall's and spectacled porpoises

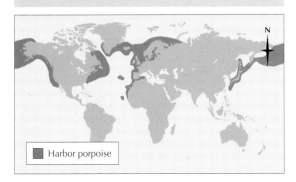

Harbor porpoise

August. Porpoises leave the Baltic Sea between November and February and if the sea freezes early, large numbers may die in the ice.

A difference in behavior between porpoises and dolphins is that porpoises are less playful. They rarely leap out of the water and do not follow boats as often as dolphins do.

Caught in nets

Because porpoises only inhabit coastal waters they are often caught and killed in fishing nets. They are also more likely to be killed by boat traffic and pollution. The vaquita, in particular, is now considered to be critically endangered, and it is estimated that there are only between 100 and 500 of these porpoises remaining in the wild. The finless porpoise, meanwhile, is now classified as being endangered while Dall's porpoise is conservation dependent.

Porpoises also fall prey to sharks and killer whales and at times are hunted by humans. They are eaten by Native Americans living on the coasts of North and South America. In Normandy there was once an important porpoise fishery. The meat was eaten and the blubber rendered into oil for lighting. Porpoise flesh was considered a royal dish in medieval England.

Nowadays porpoises are sometimes hunted because they are perceived to be a threat to commercial fisheries. In southwestern Greenland, for example, 1,000 harbor porpoises are hunted each year. The same species used to be hunted in the Baltic Sea but it has now been virtually wiped out in the region.

As with dolphins and some species of whales, porpoises, such as this harbor porpoise, are prone to stranding. However, they are more often killed by pollution or in fishing nets.

Slippery food

Porpoises feed mainly on fish, together with some squid and crustaceans. The fish they catch are pelagic (occurring in the open sea) rather than bottom-living species, which accords with the porpoises' way of life. They usually swim near the surface but have been caught in nets 200 feet (60 m) deep or more. Fish more than 1 foot (30 cm) long are rarely taken, their favorite being herring, pollack, whiting, cod and sardines. Porpoises entering estuaries are often accused of eating salmon, but there is little firm evidence for this. Dall's porpoises have been found with lantern fish in their stomachs. These are deep-sea fish that come to the surface at night.

Although porpoises have 12 to 27 teeth on the sides of each jaw, they swallow their prey whole. The spade-shaped teeth are used, rather, for gripping slippery animals. Dall's porpoise, which feeds largely on cuttlefish, has a full set of teeth, but these never show through the gums.

Suckled at the surface

Harbor porpoises mate between June and October, as do other species in the genus *Phocoena*. The gestation period lasts nearly a year, the single baby rarely being born before June. The babies, which are suckled for 8 months, are half the length of the mother when born. When suckling, the mother porpoise "blows" while lying on her side. This allows the blowhole of the baby to break the surface without it having to release the nipple. Porpoises live about 10 years.

Surf riders

Although they usually ignore boats, porpoises sometimes share with dolphins the habit of riding in the bow waves of fast-moving ships. The bow wave of a ship is caused by water being thrown out of the way, over the top of the water immediately ahead. A large, fast-moving ship sweeps a considerable wall of water in front of it. The porpoise uses this to propel itself in much the same way as a surfer on his or her board.

The trunk is raised so that the tail flukes are presented to the rush of water, which lifts the porpoise and moves it forward. Both surfer and porpoise have to keep the forepart of the body clear of the slower water in front, which would pull it down. Humans use the surfboard to provide lift at the front, whereas porpoises use their flippers.

A harbor or common porpoise, one of the more abundant species and one frequently seen off the coasts of North America.

PORT JACKSON SHARK

THE PROMINENT NOSTRILS on either side of the snout and a toothy mouth, permanently half open, give the Port Jackson shark an unattractive appearance when seen head-on. It is the best known of eight species of horn sharks. They range from 4½–6 feet (1.4–1.8 m) long and are noted for the shape of their teeth and jaws.

The Port Jackson shark, *Heterodontus portusjacksoni*, has a heavy head, blunt in front with a terminal mouth and five gill slits on each side. A ridge runs over the top of the head beside each eye. There are two dorsal fins, each with a stout spine in front, large pectoral fins and a single anal fin. The shark is brownish gray in color.

The Port Jackson shark is a common shark of the continental shelves off southern Australia, and has also been sighted off New Zealand. It is thought to have originated around South Africa. Other horn shark species, for example the horn shark (*H. francisci*), the crested bullhead shark (*H. galeatus*) and the Japanese bullhead shark (*H. japonicus*) live in the Indian Ocean and in the Pacific, around the Malay Archipelago, Japan, the Galapagos Islands and off California. They are a primitive race of sharks, related to fossil forms that date back 150 million years, to the Upper Jurassic period. The large spines associated with the fins were a feature of the forerunners of sharks and of the earliest known true sharks.

Tombstone teeth

Port Jackson sharks are fairly sluggish and live near the bottom of shallow waters, from close inshore out to about 900 feet (275 m). They tend to segregate by sex and maturity stage, and are nocturnal in habit, hiding in caves and rocky gullies during the day. At night they feed mainly on benthic (bottom-dwelling) invertebrates such as starfish and sea urchins, clams and other mollusks, and crustaceans such as crabs, which they crush with their unusual teeth.

The upper jaw fits into a deep groove and is attached to the cranium by strong ligaments. The lower jaw is slung from the cranium by the hyomandibular cartilage. The jaws themselves are lyre-shaped and the teeth have an unusual pattern. In the front half of each jaw they are very small, cone-shaped and numerous. About halfway along the jaw the teeth begin to get larger. These larger teeth are then abruptly replaced by two rows of very much larger, flattened teeth. Finally there are three or four rows

A Port Jackson shark, one of the primitive horn sharks, cruises the ocean floor in search of its bottom-dwelling prey: invertebrates such as starfish, sea urchins, clams and crabs.

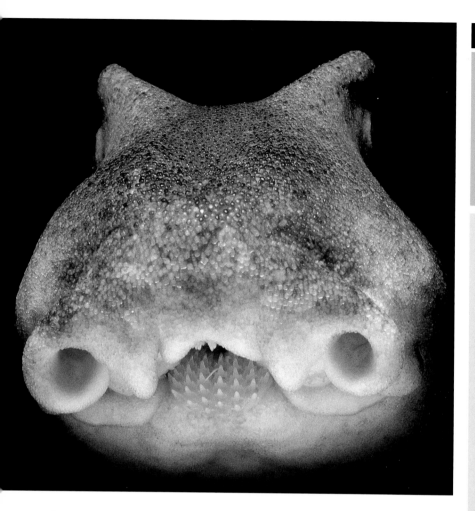

The Port Jackson shark is known for its unusual teeth and the shape of its jaws. Mainly active at night, the prominent nostrils and permanently half-open mouth give it quite a startling appearance.

of medium-sized teeth beyond these at the rear end of the jaw. It is thought that the small teeth in front of the jaw are for seizing prey and passing it backward into the mouth to be crushed by the large grinding molars.

Migrate to breed

Port Jackson sharks make yearly migrations to breeding sites. As in all sharks, the male has a pair of claspers associated with the pelvic fins that are used for transferring sperm to the female, fertilization being internal. The female lays about 10 to 16 (usually 10 to 12) eggs in rock crevices on shallow, sheltered reefs at depths of 3–16 feet (1–5 m). Egg laying usually takes place during August and September, but occasionally will start as early as July or finish as late as October. The eggs are laid over a long period of time: in captivity, females lay two eggs a day every 8–17 days. Each is in a cylindrical capsule of brown, horny material. Around the cylinder, which narrows at the lower end, are two spiral flanges.

The eggs take 9–12 months to hatch, the baby fish being 8 inches (20 cm) long when it leaves the capsule. The young then move into nursery areas in bays and estuaries. After the breeding season, males move into deeper waters, followed by the females in late September or October.

PORT JACKSON SHARK

CLASS	**Chondrichthyes**
SUBCLASS	**Elasmobranchii**
ORDER	**Heterodontiformes**
FAMILY	**Heterodontidae**
GENUS AND SPECIES	***Heterodontus portusjacksoni***

LENGTH
Up to 5⅗ ft. (1.7 m)

DISTINCTIVE FEATURES
Heavy, blunt head; prominent nostrils; toothy mouth; 5 gill slits on each side; 2 dorsal fins, each with stout spine; brownish gray in color

DIET
Bottom-living invertebrates, mainly echinoderms such as starfish and sea urchins

BREEDING
Breeding season: egg laying usually August to September; number of eggs: 10 to 12; hatching period: 9–12 months; breeding interval: 1 year

LIFE SPAN
Not known

HABITAT
Continental shelves from close inshore at intertidal zone out to 900 ft. (275 m)

DISTRIBUTION
Coasts of southern Australia and New Zealand; probably originated in South Africa

STATUS
Common

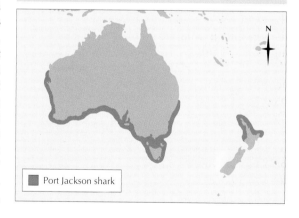

Port Jackson shark

The Port Jackson shark is considered harmless to humans, but can deliver a painful nip when provoked. The main predators of a 5-foot (1.5-m) shark such as this are likely to be other, larger sharks and commercial fishers because their flesh is said to be good to eat.

PORTUGUESE MAN-OF-WAR

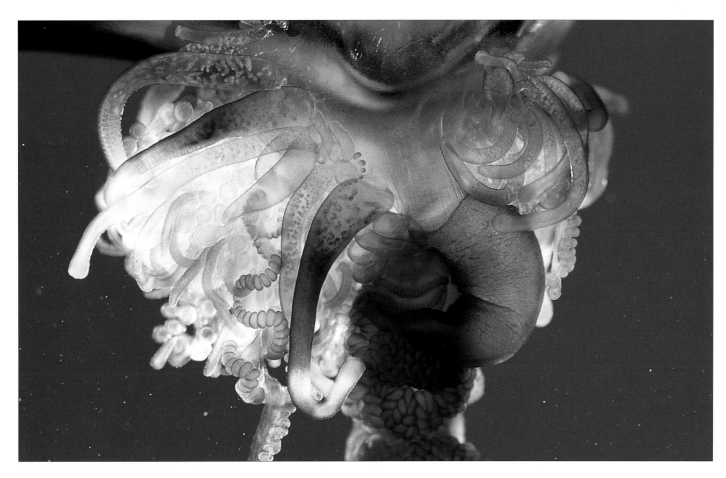

THIS COLORFUL JELLYFISH-LIKE animal, with a sting almost as powerful as a cobra's venom, floats on the surface of the sea in mid-ocean. Occasionally thousands are cast up on the coasts by strong winds. Each Portuguese man-of-war is a colony composed of four different kinds of polyps. From the water's surface, the most obvious part of the animal is a bladder-like, gas-filled, translucent float, which is in fact an enlarged polyp and may be up to 12 inches (30 cm) long, 6 inches (15 cm) high and 6 inches across. The float carries a high crest and may be blue, purple, lavender, pink or scarlet in color. Other polyps hang down from beneath the float. Some are used only for feeding while others are used in reproduction. The trailing tentacles may be 66 feet (20 m) long and are armed with nematocysts (stinging cells) to kill prey.

The most common species of Portuguese man-of-war is *Physalia physalis*. It occurs in warm seas throughout the world, especially in the North Atlantic where it normally is limited to the area covered by the circulatory currents of the Gulf Stream. It is common in tropical and subtropical areas of the Indian and Pacific Oceans, and off the coast of New South Wales, Australia.

Deadly drifter

When the wind catches the sail-like crest on the float of the Portuguese man-of-war, it causes the animal to drift across the surface of the ocean. The float can be deflated in storms, at which time the jellyfish submerges. The bladder is filled with a mixture of gases secreted by a gas gland. The gas mixture is similar to that present in air, but contains more nitrogen as well as carbon monoxide, argon and xenon. The pressure of the gas is regulated by muscles in the thin membrane forming the wall of the float. Normally these cause the float to twist and contort from time to time, so the bag dips into the water, keeping its outer surface moist. After being deflated, the float can be reinflated in a few minutes. In rough waters the tentacles are spread around the float in order to keep it balanced.

Some Portuguese men-of-war are left-sided while others are right-sided. Left-sided individuals drift at an angle of 45° to the right of the direction from which the wind is blowing, while right-sided men-of-war drift at the same angle in the opposite direction. This enables the animals to spread more evenly over the warm oceans of the world.

Commonly known as the bluebottle, Physalia utricularis *(above) is a close relative of the more numerous* P. physalis. *It is native to the Pacific and Indian Oceans.*

The Portuguese man-of-war may be caught in persistent winds blowing from one quarter over a period of weeks, and carried hundreds of miles from its normal range to be cast up in large numbers on the coasts of Europe or the United States. When this happens the bladders lie scattered in the drift line on the shore, the tentacles and polyps soon becoming shriveled and dried.

Man-of-war in action

The poison contained in the stinging cells of a Portuguese man-of-war is 75 percent as strong as a cobra's venom. When stimulated by touch or chemicals in the water each stinging cell shoots out a long tube, as in hydra (discussed elsewhere), but the surface of the tube in *Physalia* is studded with hooks, barbs and spikes of different shapes and sizes. The tube can even penetrate rubber gloves, and goes deep into the skin of small animals, remaining firmly fixed while the poison flows down the tube from the capsule of the stinging cell. The stinging tentacles, lined with batteries of nematocysts, are used for capturing planktonic crustaceans and small fish. The man-of-war shoots out the stinging cells when something touches them and also when the chemical glutathione comes into contact with the cells. This amino acid is given out from the skin of injured or dead animals.

As a fish is drawn up by the tentacles, the flask-shaped feeding polyps become agitated and move their mouths around, searching for the food. Once they taste it their mouths open wide and they fasten suckerlike onto the body of the fish. The wide mouths of a dozen feeding polyps spread over the fish, almost enclosing it, and pour digestive juices onto the prey. Smaller prey are eaten whole by a feeding polyp.

In spite of its vicious stings, the Portuguese man-of-war is eaten by several animals, in particular the loggerhead turtle, *Caretta caretta*. The latter has been seen to eat its way through a

The Portuguese man-of-war maintains its buoyancy in the water by means of a gas-filled bladder called a pneumatophore.

PORTUGUESE MAN-OF-WAR

PHYLUM	**Cnidaria**
CLASS	**Hydrozoa**
ORDER	**Siphonophora**
FAMILY	**Physalidae**
GENUS AND SPECIES	***Physalia physalis***

LENGTH
Float: usually up to 12 in. (30 cm); tentacles: up to 66 ft. (20 m) or more

DISTINCTIVE FEATURES
Blue-purple polyps; long, trailing tentacles

DIET
Small animals, including crustaceans, fish larvae and adult fish up to 4 in. (10 cm) long

BREEDING
Male and female gonophores (small reproductive sacs containing sex organs) bud. Eggs and sperm released into water, where fertilization takes place. Numerous larval phases probably take place in deep water. Breeding season: mainly autumn.

LIFE SPAN
Uncertain; estimates vary from a few months to 10 years

HABITAT
Open oceans, at surface

DISTRIBUTION
Tropical and subtropical seas and oceans; may be blown thousands of miles beyond usual range

STATUS
Common

shoal of the jellyfish, its eyes swollen and almost closed from the stings. The turtle seems to be almost immune to the stinging tentacles.

One small fish, *Nomeus albula*, shelters among the tentacles of the jellyfish and can survive 10 times the dose of poison from a Portuguese man-of-war that would kill another fish of the same size. The fish does not prey on the jellyfish, although it robs its host of some of its food.

Four kinds of polyps

In a well-grown Portuguese man-of-war there may be a thousand polyps, all of which come from one larva. One polyp from this larva forms a tiny bladder with a gas gland at one end. At the other end a tubular mouth grows out. This is the first feeding polyp. Later, an area called the budding zone appears between the bladder and this feeding polyp. The first stinging tentacle grows out from this, to be followed by more feeding polyps and stinging tentacles. By further budding the reproductive polyps are formed. The Portuguese man-of-war is unusual for a siphonophore in that it is dioecious (some individuals are male and others are female). The polyps that are used to capture prey are known as dactylozooids, while those to do with reproducing are known as gonzozooids. Those polyps that relate to feeding are known as gastrozooids.

Over time, the bladder grows into a float with a crest. It continues to grow in size, and the three kinds of polyps increase in number by budding until the mature jellyfish is formed. It probably lives for only a few months, but before dying it sheds its eggs and sperm into the sea. The eggs are fertilized and more tiny larval bladders develop from them. Reproduction takes place mainly during autumn and the spawning cycle probably begins in the Atlantic Ocean.

First mention of the species

The first written account of the Portuguese man-of-war in English seems to be in Hans Sloane's *Travels to Jamaica*, published in 1707. Sloane notes that the animal's name derives from its supposed resemblance to the shape of a Portuguese caravel, a ship with a broad bow and a high, narrow poop. Sloane's diary contains the following entry: "On Tuesday 11, when we were in about forty-six degrees of northern latitude I first saw what the seamen call a Caravel or Portuguese Man of War, which seems to be a Zoophytum, or of a middle nature between a Plant and an Animal... of a blue, purple, yellowish and white color, that burn more violently than those of the North-Sea, they do suck themselves so close to the skin that they raise Blisters, and cause sometimes St. Anthony's Fire."

Portuguese men-of-war often blow ashore after storms. Even when they have dried out, the stinging cells are dangerous and can produce a red weal on human skin similar to a severe burn.

POTOO

For the great potoo, remaining motionless during the day and impersonating the broken stump of a tree branch is an effective way of remaining undetected.

POTOOS OR WOOD NIGHTJARS are a family of seven species related to the nightjars. They have large mouths but lack the nightjars' surrounding bristles. The plumage is dark brown, streaked and mottled with buff, white and black, which makes a potoo inconspicuous. It has short legs and long wings and tail. The eyes are very large, with yellow or brown irises. Potoos live in Central and South America and in the Caribbean. The largest is the great potoo, *Nyctibius grandis*, which has scalloped upper eyelids and a bill with a 3-inch (7.5-cm) wide gape. The gray or common potoo, *N. griseus*, was once considered the most widespread species, distributed from southern Mexico to Argentina. Now, two species are identified: the gray potoos of Mexico, Central America and the Caribbean are considered distinct from those in South America, and are called northern potoos, *N. jamaicensis*.

Bright eyes at night

A potoo may often be first encountered as a pair of large yellow or brown eyes reflecting the beam of a flashlight in the dark. Otherwise potoos are usually known only by their mournful calls. Potoos are nocturnal, like their relatives. While roosting by day they remain motionless, their mottled brown plumage making them very difficult to find. Great and northern potoos live in open woodland and plantations, having favorite perches on tree stumps and branches.

Hawking for insects

Potoos hunt from their perches, keeping still until an insect flies by. Then they swoop out to catch it in the enormous gape and return to the perch. They can catch large insects such as beetles, bugs, locusts and termites. When hawking continuously for insects, potoos often return to the same perch after a feeding sally, as do owlet-nightjars (family Aegothelidae).

Hidden in the open

Potoos lay their single, white eggs in exposed positions, such as a crevice or depression in a tree stump, or the scar where a branch has fallen off. Although the nesting positions are exposed, potoos do not collect nesting material, and the

GREAT POTOO

CLASS	**Aves**
ORDER	**Caprimulgiformes**
FAMILY	**Nyctibiidae**
GENUS AND SPECIES	***Nyctibius grandis***

WEIGHT
17–22 oz (500–620 g)

LENGTH
Head to tail: 18–22 in (45–55 cm)

DISTINCTIVE FEATURES
Fairly large; slim build; big head; bill has huge gape; cryptically camouflaged, buff-brown plumage; short legs; long wings and tail

DIET
Mainly insects; possibly also small bats

BREEDING
Age at first breeding: not known; breeding season: eggs laid all year round; number of eggs: 1; incubation period: not known; fledging period: nearly 60 days; breeding interval: probably 1 year

LIFE SPAN
Not known

HABITAT
Predominantly semi-open woodland; also closed-canopy forests and plantations

DISTRIBUTION
Central America and northern South America

STATUS
Uncommon to locally common

Great potoo

The enormous gape of the great potoo is useful when trying to catch flying insects at night. Here, though, the bird is using it as a threat display.

disturbed, the potoo extends its head, with the bill pointing skyward and slightly open, the eyes being nearly closed. This makes the potoo look even less birdlike. In this posture it can be handled, when it will open its eyes wide, snap its bill and gape aggressively. Potoos also perform distraction displays, flapping and cavorting on the ground.

The male great potoo incubates the single egg during the day. The newly hatched chick is brooded, after which it perches inconspicuously, like the parents, remaining on the stump for nearly 2 months.

Weird calls

One of the features of the potoos and their relatives is their strange calls, often booming or shrieking. The potoo earned its name from a two-syllable call of the northern potoo. The same call led Haitians to call it the howling cat. Another of its cries is described as *poor-me-one*, a far-carrying, mournful whistling that, when heard at close quarters, is found to be composed of seven or eight descending notes. The great potoo produces its harsh, sickly *kwak* call at dusk and on moonlit nights. As potoos are heard at night, it is not surprising that they are connected with superstitions. To the Indians of South America, the wails of potoos are evil omens.

nests are extremely difficult for field biologists to locate. The potoo perches over the egg, covering it with its breast feathers and holding its long tail close to the trunk of the tree. In this position it is easily mistaken for a bracket fungus or the projecting stub of a broken branch. When relaxed, the potoo perches with plumage fluffed and head pulled in to its shoulders, but when

POTTO

So stealthy is the potto that it is almost impossible to see among the foliage of its rain forest habitat. A low profile helps it avoid predators and sneak up on prey.

THE POTTO IS ONE OF the lemurlike lower primates, or prosimians. A member of the Lorisidae family, it is related to the lorises (*Loris, Nycticebus*) and golden potto (*Arctocebus calabarensis*). It is 12–16 inches (30–40 cm) long and weighs 2–3 pounds (1–1.4 kg). Its 2–4-inch (5–10-cm) tail is relatively long for a member of its family. The potto has thick brown, woolly fur, rounded ears and large round, staring eyes.

The potto's digits are adapted to suit a tree-climbing lifestyle. Its thumbs and first toes are long, and its index finger and second toe are reduced (though not so much as in the golden potto), enabling hands and feet to function as clamps. Inevitably, some dexterity is lost. The snout is slender but short, adding to the whole physical impression given of a tiny bear.

On the potto's back are four or five "spines." They are in fact long extensions on the nuchal vertebrae (neck bones) and help support the shoulder and arm muscles. They also form a defensive structure over the neck, offering some protection from the bites of predators. The spines are covered with a thin layer of highly sensitive skin and sensory guard hairs.

The potto lives in the African rain forest belt from Guinea in the west to the Rift Valley in East Africa. In the Democratic Republic of Congo (formerly Zaire) it is found only in the area to the north of the Congo (Zaire) River.

Not so gentle

Living in deep tropical forest, the potto leads a solitary life, spending the day curled up almost into a ball, with the head tucked between the arms and the hands and feet clamped tightly to a branch. At dusk it comes out from its hiding place, usually a tree fork or a tangle of twigs and leaves, to move slowly and cautiously along the branches. When two pottos meet, they may groom each other and stay together for a while.

Both males and females are territorial, the latter defending just enough space to feed herself and, when present, an offspring. This is typically 15–22 acres (6–9 ha). The male defends a larger area, up to 100 acres (40 ha), which may contain the ranges of a few females but no other males. He marks the territory by smearing branches with a secretion from glands near the anus. Scent is very important to the species, imparting sexual and other information to neighboring pottos.

Slow progress

The potto dislikes crossing open spaces, preferring to keep hands and feet clasped around a branch. It moves from tree to tree only when it can do so without descending to the ground first. As it moves along a branch, the potto lets go with just one hand or foot at a time. The hand is moved forward and the fingers close around the branch in a viselike grip. Then a foot is moved to grip in the same way. This is not a very rapid method of locomotion, but the potto is not quite as slow-moving as its discoverer, Willem Bosman, declared in his description of it written in 1704. He spoke of "its lazy, sluggish nature; a whole day being enough for it to advance ten steps forward… and if the trees be high, or the way anything distant, and he meets with nothing on his journey, he inevitably dies of hunger, betwixt one tree and the other."

This is highly fanciful, as might be expected from an author who adds, "Tis impossible to look on him without horror… he hath nothing very particular but his odious ugliness." It is

POTTO

CLASS	**Mammalia**
ORDER	**Primates**
FAMILY	**Lorisidae**
GENUS AND SPECIES	***Perodicticus potto***

ALTERNATIVE NAME
Bosman's potto

WEIGHT
1⅓–3½ lb. (600–1,600 g)

LENGTH
**Head and body: 12–16 in. (30–40 cm);
tail: 2–4 in. (5–10 cm)**

DISTINCTIVE FEATURES
**Thickset body; round head; very large eyes;
opposable thumbs; short whiskers; soft,
dense coat with longer hairs on neck;
short, bushy tail; cinnamon or gray above,
darker toward head and browner near
rump; paler underbelly**

DIET
**80 percent vegetable matter, such as fruits
and leaves; also invertebrates and small
birds, mammals and lizards**

BREEDING
**Age at first breeding: 18 months; breeding
season: November–February (east of range),
April–June (west of range); gestation period:
180–205 days; number of young: usually 1;
breeding interval: 1 year**

LIFE SPAN
**Up to 11 years in wild; up to 26 years in
captivity**

HABITAT
**Primary and secondary forest edge up to
altitude of 6,560 ft. (2,000 m)**

DISTRIBUTION
Guinea east to Uganda and Kenya

STATUS
Generally common

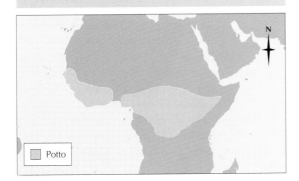

Potto

perhaps poetic justice that to distinguish it from
the golden potto this species is sometimes named
after its detractor—Bosman's potto.

Must hold on to something

Unbeknown to Bosman, pottos can move quickly
when pressed. Normally, though, they rely on
stealth, not merely to escape the notice of predators. They creep up on nesting birds so slowly as
not to be noticed and grab the victim with a
lunge and bite. This has been recorded of pottos
in Uganda, though individuals in the Ivory Coast
are mostly vegetarian. In general pottos eat fruit,
leaves, insects, snails, lizards, eggs and birds.

The female comes into season about every 40
days. The infant (usually just one) clings tightly
to its mother's belly, snuggling against her for
warmth. An offspring grows rapidly; its spines
begin to appear at 30 days. When the mother is
moving around in the trees, she often parks her
baby, leaving it clinging to a branch, where it
will be quite safe until she comes to fetch it. The
youngster is weaned at 4 or 5 months and
reaches adult size at 8 to 14 months. The record
life span in captivity is 26 years.

Spines toward the enemy

The potto's natural predators include civets,
leopards, owls and snakes such as pythons. The
potto's defense response is to curl up, clinging
with its hands and feet, head tucked between its
arms as in sleep, turning its spines toward the
enemy. If threatened further, the potto suddenly
releases the branch with its hands and lashes out
with them. At the same time, it can inflict painful
bites with its small, sharp teeth. In some parts of
its range people hunt it for food, and it is also
under increasing pressure from habitat loss.

*Whenever possible, the
potto keeps three points
of contact on its perch.
If threatened, it may
escape by rolling into a
furry ball and dropping
to the forest floor.*

PRAIRIE CHICKEN

The male lesser prairie chicken, using his characteristic display to attract females, at a booming ground.

THE TWO SPECIES OF prairie chickens are grouse that live in the central United States. The greater prairie chicken is 16–18½ inches (41–47 cm) long. It is pale brown with black barring and the face is white with a black eyestripe. The male has an orange crown, large orange air sacs and an epaulet of stiff, pointed feathers around the neck that are raised during display. The lesser prairie chicken is similar in appearance, but it is slightly smaller and has red or violet air sacs. The feet are slightly feathered to form snowshoes.

Before the spread of European civilization, prairie chickens were widespread over the grasslands of the United States, from the Canadian border to Texas and eastward to the Atlantic. The main center of their population was the previously vast prairies of the United States, but they also flourished in open woodlands where there were grassy clearings. The lesser prairie chicken lives in country dominated by dwarf oak bush. They are now very much restricted. Radical alterations of the habitat by humans, and predation by domestic dogs and cats are the main causes.

Habitat destroyed

When Europeans first spread westward in North America, cutting the woodlands down for pasture, the prairie chicken flourished on a diet of grass grains. Later the prairies were plowed up, robbing the prairie chickens of their homes. This process was accelerated during and after World War I, when grain growing was a strategic necessity and the development of mechanization allowed large areas to be cultivated. The prairie chicken became extinct in the eastern part of its range and only survived in isolated pockets. Yet more advanced farm machinery threatens some of these pockets, but others have been set aside as reserves.

One race of prairie chicken has already become extinct. This is the heath hen, a subspecies of the greater prairie chicken, which used to live along the North Atlantic seaboard in what were probably grasslands that sprung up after forest fires. The heath hen disappeared from the mainland of New England in about 1835 but survived on Martha's Vineyard until 1932. Meanwhile, the population of Attwater's subspecies, found in Texas, has dwindled to double figures.

Booming grounds

Like other grouse, male prairie chickens meet in communal display sites called booming grounds, which are examples of a wider phenomenon in the animal kingdom: the use of leks. The booming grounds are used from one year to another. The males gather in early spring, and every day for several months, at dawn and dusk, they indulge in their displays. The orange air sacs are inflated and the neck epaulet and tail are raised. Once arrayed, the males scurry to and fro and spin in circles with head held low, wings drooping and tail alternately fanned and shut with a loud click. While dancing, the prairie chickens utter a three-syllable boom that, in chorus, can be heard as a continuous humming for up to 4 miles (6.5 km); a typical distance between neighboring booming grounds.

Each male prairie chicken stakes out a territory within the booming ground, and competes with its neighbors to occupy the largest territory. The booming ground acts as an arena for mate choice by the females. The females visit the booming ground to mate and then depart to rear the family by themselves. The most successful males are those with the largest territories. On a booming ground that was watched for 2 years, there were nine males. Each year, one of the nine males dominated and held a large territory.

PRAIRIE CHICKENS

CLASS **Aves**

ORDER **Galliformes**

FAMILY **Tetraonidae**

GENUS AND SPECIES **Greater prairie chicken,** *Tympanuchus cupido;* **lesser prairie chicken,** *T. pallidicinctus*

WEIGHT
Greater prairie chicken, male: about 2⅛ lb. (990 g); female: 1¾ lb. (770 g)

LENGTH
Greater prairie chicken, head to tail: 16–18½ in. (41–47 cm)

DISTINCTIVE FEATURES
Greater prairie chicken. Very plump, medium-sized bird; buff throat; brown and buff bars on body; short, rounded tail.

DIET
Grain, leaves, seeds, buds and insects

BREEDING
Age at first breeding: 2 years (male), 1 year (female); breeding season: eggs laid April to June; number of eggs: 8 to 13; incubation period: 23–25 days; fledging period: 100 days

LIFE SPAN
Usually up to 3 years

HABITAT
Savanna, scrub, open bush and forest

DISTRIBUTION
Midwestern U.S.

STATUS
Scarce; Attwater's prairie chicken (Texas race of greater prairie chicken): threatened

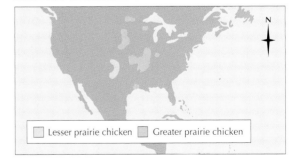

Lesser prairie chicken Greater prairie chicken

This male also took part in about 70 percent of the matings, while the weakest males did not mate at all. In this way, most prairie chicken offspring are fathered by the male that is most successful at competing on the booming grounds, a clear example of selection of the fittest.

The female prairie chicken makes a nest lined with feathers, dry grasses, leaves and twigs, and sites it within thick grass cover. The chicks leave the nest shortly after hatching and follow their mother. They are very precocious and are already able to catch their own prey at this stage. They fledge in 14 weeks and are independent in about 10–12 weeks.

Safe in the open

In the course of another study on prairie chickens, booming grounds were watched for 4,745 mornings over 24 years. Among other observations, a watch was kept for interactions between predators and the prairie chickens. The results were surprising. In all that time only four prairie chickens were killed on the booming grounds. Three were killed by birds, a harrier, a snowy owl and a horned owl, and one by a mongrel dog from a nearby farm. Although foxes and coyotes visited the booming grounds, none made a kill. The prairie chickens were more wary of dogs than of foxes or coyotes. The reaction to a ground predator was surprisingly casual. As the predator approached, the prairie chicken walked out of the way with neck stretched up. If pressed it took flight, but settled within 100–200 feet (30–60 m). Sometimes foxes were seen to cross the booming ground, taking no notice of the prairie chickens, which similarly disregarded the foxes.

Predation on the booming grounds is far less successful than one might expect. Perhaps it is easy for the birds to stay watchful when out in the open and in a group. At other times, for instance in the fall, when dispersal of juveniles takes place, mortality can be very high. In fact, over 50 percent of a prairie chicken population dies every year, and life expectancy in the wild is no more than 2–3 years.

The bright colors and conspicuous display of the male greater prairie chicken are intended to reach the female of the species, but also catch the eye of predators. However, many more are killed as juveniles than during mating displays.

PRAIRIE DOG

PRAIRIE DOGS ARE hamsterlike, short-tailed ground squirrels, so named because of their barking calls. Apart from being slightly flattened, the prairie dog's tail bears little resemblance to that of a squirrel, although its head is more squirrel-like, apart from its small ears. The length of the head and body averages about 1 foot (30 cm) with a 3½-inch (9-cm) tail. The prairie dog's fur is yellowish brown, gray or slate gray with paler cream or white underparts. In some species the tip of the tail is black. All five species are very similar in appearance.

Prairie dogs inhabit grassland and pasture on the plains and plateaus of North America, from the Dakotas to Texas, and from Utah and Arizona in the west to Kansas and Oklahoma in the east. They are also found in southern Canada and northern Mexico. The black-tailed prairie dog, *Cynomys ludovicianus*, is the most wide-ranging throughout this area. The other four species occur more locally in the southern part of this range.

Prairie citizens

Prairie dogs used to live in vast populations, called towns. In 1901 one prairie dog town was estimated to cover an area 100 by 240 miles (160 by 390 km) and to contain 400 million animals. Nowadays towns are much smaller, because large concentrations of prairie dogs inevitably came into conflict with humans by competing for land with grazing cattle.

As with other colonial or social animals, prairie dogs have a social organization, which, as must be expected with such vast colonies, is very complex. A single town is divided into a number of wards, the boundaries of which depend largely on the geography of the area and linear features such as streams. The wards are then divided into a number of coteries (groups of individuals), each covering less than 1 acre (0.4 ha). The coterie is the base unit on which each prairie dog's life is founded. Each family unit defends its territory, and individuals rarely venture outside the boundaries. If they do, they are likely to be chased back by members of neighboring coteries.

A typical coterie consists of an adult male, three adult females and a variable number of young prairie dogs. The members of a coterie recognize each other and are on friendly terms and, except for the very young ones, they jealously guard the coterie's boundaries. Apart from squabbles along the borders, members of the coterie, including the youngsters, advertise their territories with a display. Each animal rears up on its hind legs, with its nose pointing to the sky, then delivers a series of two-syllable calls.

Each coterie has a network of burrows with a large number of entrances. From the entrance the burrow descends steeply for 3–4 yards (2.7–3.6 m) before meeting radial tunnels with nests at the end. From a distance a prairie dog town appears pockmarked with craters because each burrow entrance is surrounded by a volcano-like cone. This is more than the casual accumulation of excavated soil:

A Gunnison's prairie dog, Cynomys gunnisoni, *Bruce Canyon National Park, Utah. Prairie dogs used to be found in vast towns across the plains of North America but are now mostly confined to National Parks.*

PRAIRIE DOGS

CLASS **Mammalia**

ORDER **Rodentia**

FAMILY **Sciuridae**

GENUS AND SPECIES **Black-tailed prairie dog,** *Cynomys ludovicianus*; **white-tailed prairie dog,** *C. leucurus*; **Gunnison's prairie dog,** *C. gunnisoni*; **Utah prairie dog,** *C. parvidens*; **Mexican prairie dog,** *C. mexicanus*

WEIGHT
1½–3 lb. (0.7–1.4 kg)

LENGTH
Head and body: 11–13 in. (28–33 cm); tail: 1–4½ in. (3–11.5 cm)

DISTINCTIVE FEATURES
Hamsterlike appearance; squirrel-like head with small ears; yellow-brown, gray or slate gray upperparts; paler cream or white underparts; flattened, often black-tipped, tail

DIET
Grass seeds and shoots

BREEDING
Age at first breeding: 18–24 months; breeding season: January–May (southern U.S.), February–April (northern U.S.); number of young: usually 3 or 4; gestation period: 27–35 days; breeding interval: 1 year

LIFE SPAN
Up to 8 years in captivity

HABITAT
Open grassland and pasture

DISTRIBUTION
Southern Canada; western U.S. south to northern Mexico

STATUS
Some species locally common; conservation dependent: Utah prairie dog; endangered: Mexican prairie dog

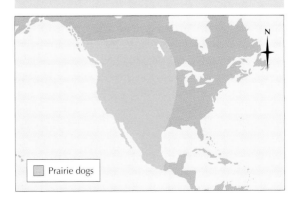

Prairie dogs

it is actually a carefully built rampart of soil 1–2 feet (30–60 cm) high and up to 6 feet (1.8 m) across. The soil is gathered from the surface, brought to the entrance and patted into place, where it serves as a lookout post and a protection against floods after heavy rain.

Unintentional agriculture

Prairie dogs are vegetarian, feeding on grass shoots and seeds, and other plants that grow on the prairies. Not surprisingly, the crowds of prairie dogs have a profound effect on the vegetation inside the town limits. Taller plants are completely eliminated, being cut down and left to wither if they are not eaten. Meanwhile, the continual cropping of the grasses and herbs encourages fast-growing plants with abundant seeds and also encourages high herbaceous plant diversity. As a result, the optimal vegetation, from the prairie dog's point of view, is produced. There is a second advantage to this unintentional agriculture. The removal of tall plants deprives predators of cover and allows the prairie dogs a clear view from their mounds.

Prairie dogs, such as this black-tailed prairie dog, are named for their barking alarm call. This is specialized to defend against different predators.

Prairie dogs live in close-knit family groups known as coteries. When food is plentiful, a female will suckle any pup within the coterie.

Perhaps the main advantage of the coterie system is that each group of prairie dogs has sufficient area for feeding, and overgrazing is prevented by not allowing other prairie dogs onto the pasture. When the population gets too large, some members of the coterie emigrate.

Keeping the balance

The rate of reproduction in prairie dogs is slow compared with that of many other rodents. Each female produces only one litter each year, usually of three or four pups, although there may be up to eight young. Breeding takes place from January to May in the southern United States and between February and April in the north. The young are born after a gestation period of 27–35 days. The pups' eyes open at 33 days and they are weaned in 7 weeks.

Although comparatively few pups are born each year, the population can still increase rapidly, for example, growing from 4 to 15 prairie dogs per acre in 3 months. Such an increase would threaten the food supply if it were not for emigrations to "overspill" towns or suburbs. When the population rises, the behavior of the prairie dogs changes. Usually any member of a coterie can enter any burrow, and any female will suckle any pup. However, when the population becomes too high, the females defend their nests. The other adults and last year's young, meanwhile, dig burrows and feed at the edge of the town, commuting home at night. As the

young prairie dogs appear, the traveling animals move permanently into their new homes. The population is thus redistributed without disturbing the boundaries.

Early warning system

Prairie dogs fall prey to many predators, particularly coyotes and raptors (birds of prey). As with marmots (discussed elsewhere in the encyclopedia), prairie dogs have an alarm call that sends them all bolting for cover. This is the bark that is responsible for their name. It is a short nasal yip and is specialized for different predators, such as the black footed ferret. The alarm call also has several shades of meaning. When high-pitched, it is the signal for immediate flight. The territorial call, however, is used as an all-clear signal.

The depredations of so many predators does not seriously affect prairie dog numbers, but the prairie dog's use of the grassy plains has led to its being nearly wiped out by humans. Ranchers want the grasses for their livestock, which are also in danger of breaking their legs in prairie dog burrows. In the past, poisoning has been so effective that prairie dog towns have been wiped out across much of their former range and they now survive mainly in national parks and other protected areas. While some species are still locally common, with prairie towns of up to several thousand animals in places, the Utah prairie dog, *C. parvidens,* is now conservation dependent and *C. mexicanus* is endangered.

PRATINCOLE

THE NAME PRATINCOLE is derived from the Latin for "meadow-dweller," but the rarely used alternative name of swallow plover is perhaps more appropriate for these relatives of the coursers (discussed elsewhere). In flight pratincoles look like large swallows, with long pointed wings and forked tails. Although their legs are short, they run well, holding their bodies parallel to the ground. The middle and outer toes are connected by a web of skin, and the claw of the middle toe is very long, with a comb, as in herons. The Australian pratincole, *Stiltia isabella*, lacks this comb. The bill of a pratincole is short, broad at the base and pointed at the tip.

The plumage of pratincoles is generally dull brown, with white underparts and a white rump. The most plentiful species, the common, collared or red-winged pratincole, *Glareola pratincola*, has a buff throat edged with black. The base of the bill is shiny red and the rest is black. The black-winged pratincole, *G. nordmanni*, differs only in having entirely black underwings.

The eight species of pratincoles live in Africa, Asia, Australia and southern Europe. The common pratincole and black-winged pratincole occasionally wander into western and central Europe, including the British Isles.

Mass landings

Pratincoles live in open and often very desolate country, on the edges of marshes, estuaries, plains and deserts. They have a very buoyant flight and, with their long wings and forked tails, look very much like terns or jaegers. Pratincoles live in flocks, usually near water, and feed together, flying to and fro after insects in a dense cloud. Especially in the evening, members of the flock call to each other with a pleasant trill-like song. When a flock settles, pratincoles land packed together, all facing in the same direction. As they alight, each bird stands for a few moments with its wings held stretched above its back. When they have folded their wings the pratincoles again remain motionless, and even a large flock is difficult to see.

Some pratincoles are migratory, the common pratincole moving from Asia to Africa from September to March. Others make local movements in search of suitable living conditions.

Flying insect nets

Pratincoles live on insects, particularly grasshoppers, crickets, locusts and many types of beetles. Flies, dragonflies, bugs and others are also taken. In Africa pratincoles are called locust birds, from their habit of following swarms of locusts. One black-winged pratincole was found to have 135 locusts in its crop.

Pratincoles catch some insects on the ground. Typically, however, they hawk for insects in the air. As they zigzag to and fro, their short, triangular bills open to expose a wide gape in which the insects are quickly netted.

Dancing pratincoles

Pratincoles breed on small islands in lakes, on sandbanks in rivers, on mudflats or in deserts where their nests are fairly safe from predators. These sites usually are not permanent, and clutches may be destroyed by floods.

Breeding grounds also vary from year to year. Several hundred pairs of pratincoles may nest together in loose colonies, each pair staking out a territory around the nest. Intruding neighbors are threatened by lowering the breast right to the ground, with wings spread or half held out and tail raised.

Sometimes communal displays take place on neutral ground. Several pairs of pratincoles gather and take part in what looks like a square

A common pratincole in Chobe National Park, Botswana. Pratincoles often live near water, along the edges of marshes, pools and estuaries.

The Australian pratincole does not have a comb on the claw of the middle toe, which is characteristic of other pratincoles.

dance, each couple bowing to the other with tails raised, bobbing their heads and even leaning gently against one another.

Nesting usually takes place after the rainy season. The nest is no more than a scrape in the ground, and sometimes a hoofprint or a dry cowpat is used. Two or three, sometimes four, eggs are laid and are incubated by both parents. During the day the efficiently camouflaged eggs may be left to be warmed by the sun, and when it is very hot the parents may shade the eggs or chicks with their outstretched wings.

The chicks, which hatch out after 17–19 days, are at first fed by the parents and begin to feed themselves when just 2 weeks old. They can run about almost immediately, and start to fly when approximately 3–4 weeks old.

Pretended massacre

When approached on the nest, pratincoles sit tight to protect their eggs or chicks. Then, when dislodged, they fly to and fro, calling. They also resort to distraction displays, which are all the more impressive because a number of pratincoles display at the same time.

They display by pretending to have broken wings, and descriptions have been given of mass displays in which 20 or so pratincoles fluttered to the ground, thrashing about with one or both wings trailing, presenting a pitiable sight as if a shotgun had been discharged into the flock. No doubt the sight is sufficient to distract any predator from looking for the eggs and chicks.

COMMON PRATINCOLE

CLASS	**Aves**
ORDER	**Charadriiformes**
FAMILY	**Glareolidae**
GENUS AND SPECIES	***Glareola pratincola***

ALTERNATIVE NAMES
Collared pratincole; red-winged pratincole; locust bird (Africa only)

WEIGHT
Male: usually 2½–3½ oz. (70–95 g)

LENGTH
Head to tail: about 10 in. (25 cm)

DISTINCTIVE FEATURES
Slender body; small bill with broad base and pointed tip; very long wings; short legs; web of skin between middle and outer toes; deeply forked tail. Breeding adult: buff throat with black necklace; olive-brown head and upperparts; white rump and belly.

DIET
Insects such as grasshoppers, crickets, locusts and beetles

BREEDING
Age at first breeding: 1 year; breeding season: eggs laid late April–May (Europe); number of eggs: 3; incubation period: 17–19 days; fledging period: 25–30 days; breeding interval: about 1 year

LIFE SPAN
Not known

HABITAT
Flat, open country without trees, shrubs or tall vegetation; often near water

DISTRIBUTION
Scattered range in Africa and from southern Europe east to Kazakhstan and Pakistan

STATUS
Locally common

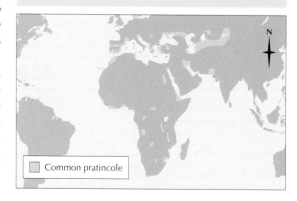

Common pratincole

PRAWN

EVEN THOUGH THE TERM *prawn* has little scientific standing, scientists still use it. It originally referred to one species that lives around the coasts of Europe, and this usage continued for 500 years or more. Then gradually, as more and more marine animals were studied, the name was applied to all sorts of long-bodied, small crustaceans. In the United States, most of these would be called shrimps, although the name prawn is often applied to some larger species and to some freshwater species.

The original prawn, *Palaemon serratus*, often called *Leander serratus*, is a large, shrimplike crustacean, about 4⅓ inches (11 cm) long when straightened out. A saber-shaped projection (the rostrum) on its head shield is toothed or serrated, and the eyes are stalked. The outer skeleton is thin and hornlike in texture, and there are two pairs of very long, filamentous antennae. The first pair has three branches, one extending forward, the other two upward and backward. The antennae of the second pair are undivided. The prawn has five pairs of limbs on the thorax, the first and second pairs having small but distinct pincerlike claws. Five pairs of small swimming legs line the narrow abdomen, which ends in a fanlike tail.

Palaemon serratus belongs to the family Palaemonidae, which also includes among its members *Palaemon elegans*, another European species, together with the red-handed shrimp (*Palaemon serenus*) of southeastern Australia, the African prawn (*Palaemon africanus*) and the Oriental shrimp (*Palaemon macrodactylus*). The freshwater prawns of the genus *Macrobrachium* also belong to the Palaemonidae. Among these are the giant freshwater prawn of the United States (*Macrobrachium carcinus*), which weighs up to 3 pounds (1.4 kg), and the Malaysian prawn (*Macrobrachium rosenbergii*), which is a native of Southeast Asia. Both the African prawn and the Oriental shrimp, as well as several *Macrobrachium* prawns, have been introduced into the United States.

Three ways of moving

Prawns of the genus *Palaemon* live inshore, on sandy bottoms over which they walk gently forward on the rear three pairs of thoracic limbs. They can also swim gently, using the abdominal swimming legs, or shoot suddenly backward by flexing the abdomen sharply under the thorax. This style of swimming is typical of all shrimplike decapod crustaceans. Like so many other crustaceans, including crabs, prawns are scavengers, eating any small pieces of dead flesh that they find. They probably also eat small fragments of various kinds of seaweed.

Because of its external skeleton, or shell, which holds the body like a straitjacket, a prawn, like all crustaceans, cannot gradually grow in size but must do so in stages, with a molt at each stage. As is the case in other animals, growth is more rapid when a crustacean is young and slows down with age. In crustaceans, the period between molts may be days, weeks or even months, the time varying with the species and with the individual's age.

Having fed for a while and stored enough energy to grow to a larger size, a prawn grows a new shell under the old one. In a mature prawn this takes about 2 weeks. The old shell then splits across the back, at the point where the abdomen joins the thorax. The prawn pulls its legs out of its old covering, drags its abdomen free, and then gives a sudden upward jump out of its old shell, which lies empty on the sand, like a ghost. While the new shell is still soft, the prawn takes in water, which makes the body swell up, stretching the new shell. The shell later

Palaemon serratus is a large prawn found in British waters. It is usually associated with rock pools, where it lurks among weeds.

hardens while the prawn's body returns to its normal size. In this way, the prawn creates room inside the new shell for more growth. It takes just 10–20 seconds for the prawn to shed the old shell, but it takes 2 days for the new shell to harden. The prawn hides during this period because it is very vulnerable to predators.

Mating while molting

For female prawns, molting time is also mating time. In early summer, when her ovary is ripe, the female sheds her old shell. While her new shell is still soft, the male pairs with her, placing his sperm on her underside near the bases of the third pair of thoracic limbs. The two ducts from the ovary open here, and the female, resting on her fan-shaped tail and on the tips of her thoracic limbs, arches her body and then lays her eggs. These are fertilized by the male's sperm as they stream out and the female sweeps them back under the abdomen using the first pair of swimming legs. The other swimming legs, except for the last pair, secrete a cementlike substance from glands near their bases, which glues the eggs in a mass to bristles on the female's underside. Although spawning takes only a few minutes, getting the eggs in position and fixing them takes an hour. When carrying eggs, the female is said to be in berry. Some days later the eggs hatch and the tiny larvae swim away, becoming part of the plankton for a while before changing into young prawns.

The term prawn *has little scientific standing, and the division between prawns and shrimps is arbitrary. The species pictured, known as the painted harlequin prawn, occurs in the Pacific Ocean off Hawaii.*

Many predators

One of the prawn's main predators is the cuttlefish, although almost any bottom-feeding animal is likely to include prawns in its diet. Prawns are also highly palatable to people. They are in such great demand that huge areas of tropical coastline have been devoted to their cultivation. This industry has led to large-scale destruction of coastal habitats such as mangrove swamps to create prawn ponds that are often abandoned after a few years. For centuries people have also netted prawns for food. Trawling of the seabed for wild prawns and shrimps is not without its problems, since the by-catch (unwanted, often very small, fish and other animals) is often greater than the prawn and shrimp catch. Much of the by-catch dies before or shortly after being dumped back overboard.

PRAWN	
PHYLUM	**Arthropoda**
CLASS	**Crustacea**
ORDER	**Decapoda**
FAMILY	**Palaemonidae**
GENUS AND SPECIES	***Palaemon serratus***

LENGTH
Up to 4⅓ in. (11 cm) when straightened out (excluding antennae)

DISTINCTIVE FEATURES
Transparent; toothed rostrum (saber-shaped projection on top of head); stalked eyes; 2 pairs of very long antennae; 5 pairs of walking legs; fanlike tail

DIET
Wide variety of small animals, seaweed and organic detritus

BREEDING
Breeding period: during female molt; eggs carried beneath female's abdomen until hatching time

LIFE SPAN
Probably about 2–5 years

HABITAT
Inshore waters with sandy seabeds, including rock pools

DISTRIBUTION
European coasts

STATUS
Abundant

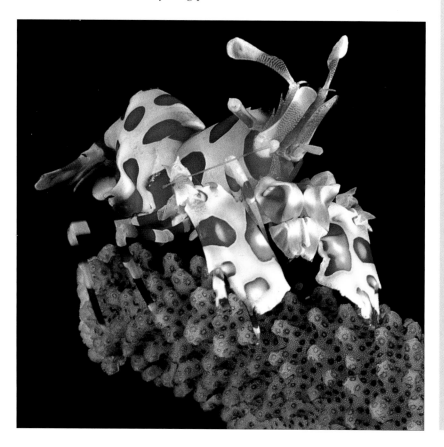

PRION

Prions are birds of the open ocean, coming to land only to breed. Small relatives of the fulmar (described separately), there are six species, the broad-billed, Salvin's, Antarctic, slender-billed, fairy and fulmar prions. They all live in the southern oceans. They are slate gray above and white underneath, with a more or less prominent white stripe above the eye and a dark patch just below the eye in most species. The upper surfaces of their wings are darker than the body, with a black W visible when prions are in flight. The bill has the tubular nostril typical of the petrel and albatross families, and in two species is fringed on each side by a row of lamellae (thin protrusions), like a short-toothed comb. The tail is wedge-shaped. The legs and webbed feet are pale blue.

The largest prion is the broad-billed prion, *Pachyptila vittata*, at 10–12 inches (25–30 cm) long. The bill is black or gray, and is very broad at the base. The thin-billed prion, *P. belcheri*, is slightly smaller at 10 inches in length, but its bill has a much narrower base, giving it a very different shape. At sea, prions are easily confused with the blue petrel, *Halobaena caerulea*. The prions have wedge-shaped tails with black edges to the central feathers, whereas the blue petrel has a square tail edged with white.

Prions breed from the coasts of Australia and New Zealand to Cape Denison on the mainland of Antarctica due south of New Zealand. They are also found on the islands surrounding Antarctica and as far north as Tristan da Cunha and Chatham Island. Outside the breeding season they fly northward, occasionally reaching Madeira and Indonesia.

Ocean flocks

Prions travel in flocks over the ocean, flying swiftly and flicking from side to side, exposing their white bellies. They are often seen from ships in southern seas, but they appear to be less common than they really are because their dull plumage blends so well with the sea, especially in Antarctic seas where visibility is often poor. The nonbreeding season is usually spent well away from land.

Filter feeding

Prions congregate to feed where upwellings or mixing of the seawater cause an abundance of planktonic animals and plants. The various species of prions appear to feed in different places, each exploiting a different source of food, which prevents competition.

A nesting pair of fairy prions, Pachyptila turtur, *sometimes has to share its burrow with another pair.*

Prions pick food out of the sea while they creep into the wind, their bodies resting lightly on the water surface. They pedal continually with their feet, disappearing and reappearing, while scooping up small organisms.

Two species of prions are known to sailors as whalebirds because the lamellae on the bill act as strainers in the same way as a whale's baleen plates. They take in a mouthful of water, then squeeze it out between the lamellae, leaving any plankton behind. The shape of the bill seems to be connected to feeding technique. In New Zealand waters, for instance, the broad-billed prion feeds on very small organisms, mainly tiny crustaceans in summer and small squid in winter. The lower part of the bill is pouchlike and drops down, letting in a large mouthful of water, like the lowering of the floor of the mouth by a blue whale, *Balaenoptera musculus*. The narrow-billed prion, on the other hand, feeds on crustaceans up to ½ inch (12 mm) long, which can easily be taken with its narrow bill.

Nesting in burrows

Prions nest in burrows that they dig in the soil, under tussock grass or in deep banks of mossy peat. Where the soil is scant or missing, they nest in crevices or under boulders. The burrows, up to 6 feet (1.8 m) long, are dug with the bill, the feet being used to kick the loose soil out.

Breeding season is the only time these birds may be found on land. This broad-billed prion was photographed in New Zealand.

BROAD-BILLED PRION

CLASS	**Aves**
ORDER	**Procellariiformes**
FAMILY	**Procellariidae**
GENUS AND SPECIES	***Pachyptila vittata***

LENGTH
Head to tail: 10–12 in. (25–30 cm)

DISTINCTIVE FEATURES
Blue-gray upperparts with black W on upperwings, obvious in flight; thick, black bill; long wings; wedge-shaped tail

DIET
Mainly small crustaceans, squid and fish

BREEDING
Age at first breeding: probably 5–6 years; breeding season: eggs laid July–August; number of eggs: 1; incubation period: about 50 days; breeding interval: 1 year

LIFE SPAN
Not known

HABITAT
Breeds on coastal slopes and lava fields, cliffs and offshore islets; outside of breeding season, rarely visits land

DISTRIBUTION
Breeding: South Island, New Zealand; Chatham Island; Tristan da Cunha; Gough Island; rest of year: southern oceans

STATUS
Common; abundant on Gough Island

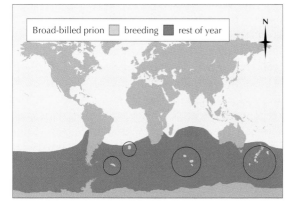

Broad-billed prion ☐ breeding ■ rest of year

Nesting begins in the spring when the prions start visiting their old burrows or digging new burrows at night. At this time the air around the cliffs and slopes where they breed is filled with a continuous chorus of dovelike cooing as courtship takes place. The nests are usually found in colonies, some of them huge. There are an estimated 8 million pairs nesting on Crozet Island in the Indian Ocean.

After mating and digging, the females leave the burrows for a few days for intensive feeding, extra food being needed to build up the single large egg. The egg is usually laid the night they return from feeding and is incubated for about 50 days in the broad-billed prion, or 43–52 days in Salvin's prion, *Pachyptila salvini*, each parent sitting for 3–4 days at a time. The chick starts life as a ball of light gray down and is brooded for 3–5 days. When it is hungry, it nibbles a parent's breast feathers and calls in high-pitched cheeps. The parent then regurgitates food and the chick puts its head in the parent's mouth to take it. The chicks leave the nest when about 50 days old.

Dug out or frozen in

The main predators of prions are the skuas, the large, gull-like predators that feed on penguins and other seabirds. It is unlikely that skuas catch many prions in flight, but wait for them as they land and shuffle toward their burrows. Skuas also dig open the burrows to get at adults or chicks, and the ground around a skua's nest may be littered with bones and feathers, the remains of prions and other birds.

In the Antarctic prion, *Pachyptila desolata*, eggs and chicks are often endangered by the hostile weather. If a snowstorm covers the burrow, the parents can dig down to it, unless repeated thawing and freezing have left a hard crust over the entrance. Some burrows are liable to be filled with drift snow or melted water, which kill both eggs and chicks by freezing.

In common with other birds that lay only one egg, prions tend to be relatively long-lived, and their reproductive cycle is slow. Sexual maturity is not reached until the birds are several years old. This means that if a population suffers a crisis, recovery of the population is slow.

PROCESSIONARY MOTH

ALSO KNOWN AS processional or procession moths, these insects attract the limelight more as caterpillars than as adults, because as larvae they form long processions, walking in follow-the-leader formation for considerable distances. Usually processionary moths are classified in their own family, the Thaumetopoeidae, but they are occasionally placed in the Notodontidae. They live Eurasia, North Africa and Australia.

Two species of processionary moths occur in southern Europe: the oak processionary moth, *Thaumetopoea processionea*, and the pine processionary moth, *T. pityocampa*. The male of the oak processionary has a wingspan of just over 1 inch (2.5 cm). Its forewings are grayish white with dark bars and the hind wings gray with indistinct dark bars. The female is 1¼ inches (3 cm) across the wings and is lighter in color. Both sexes rest on bark during the day with the wings folded, and fly among the tops of oaks at night from August to September.

Communal webs

Female oak processionaries lay their eggs on the trunks of oaks, and the caterpillars hatch late in the following spring. They are dark bluish gray, whitish on the sides with white hairs and have rows of rusty red hairy warts on the back. They grow to 1¾ inches (4.5 cm) in length. The caterpillars spin a communal web, usually at the base of a tree, and emerge at night in a procession to feed on the oak leaves. The larvae pupate at the end of June in the communal web, the brownish pupae forming tight groups.

The pine processionary moth is similar in appearance, but is 1½ inches (3.8 cm) across and is on the wing from May to July. The eggs are laid in columns on pines and the mass is covered with hairs from a tuft located on the adult female's abdomen. The caterpillars are also 1½ inches long, grayish green with tufts of white hairs, a dark back stripe and yellowish red warts. Pine processionary caterpillars cluster by day but have no permanent communal nest, and they pupate individually, deep in sandy soil. They can be serious pests in coniferous forests.

Round and round interminably

The habits of processionary moths are best illustrated by a story told by the French entomologist J. H. Fabre. He persuaded the leader of a line of caterpillars to walk onto the rim of a tub 5 feet (1.5 m) in circumference. When enough caterpillars were on the rim to form an unbroken circle,

he removed the rest and cleaned away the trail of silk they had left behind. With the head of each touching the rear end of the one in front, the caterpillars steadily marched around and around the rim, continuing for 8 days until some of them, either from exhaustion or accident, fell from the rim and the spell was broken. The caterpillars circled the rim of the tub 335 times, stopping at intervals for a rest.

Two kinds of processions

Pine processionary caterpillars travel in single file, and as many as 300 have been counted, in a line stretching 40 feet (12 m). Caterpillars of the

Processionary nests are large tents of silk spun on the food plant by the caterpillars. Pine processionaries can be so common that there are several nests on a single tree.

Pine processionary caterpillars leave the nest in line to go in search of food. When the time comes to pupate, they also set out in line before settling down singly in the soil.

PROCESSIONARY MOTHS

PHYLUM **Arthropoda**

CLASS **Insecta**

ORDER **Lepidoptera**

FAMILY **Thaumetopoeidae**

GENUS AND SPECIES **Australian processionary moth,** *Ochrogaster lunifer*; **pine processionary moth,** *Thaumetopoea pityocampa*; **oak processionary moth,** *T. processionea*; **pistachio processionary,** *T. solitaria*; **others**

LENGTH
Thaumetopoea. **Adult wingspan: 1¼–2 in. (3–5 cm). Caterpillar: about 1½ in. (3.8 cm).**

DISTINCTIVE FEATURES
Thaumetopoea. **Adult: drab coloration of grayish white forewings banded with brown. Caterpillar: generally grayish with white hairs and reddish warts.**

DIET
Caterpillar feeds on leaves of host plant. Australian processionary: wattle leaves. Pine processionary: pine needles. Oak processionary: oak leaves. Pistachio processionary: pistachio leaves.

BREEDING
Varies according to species

LIFE SPAN
Thaumetopoea: **slightly less than 1 year**

HABITAT
Woodland containing host species

DISTRIBUTION
Eurasia, North Africa and Australia

STATUS
Some species abundant

oak processionary moth travel in generally shorter, less ordered groups. Each processionary caterpillar lays a thread of silk as it goes, so the trail is marked by a substantial carpet of silk. This was the trail Fabre removed at the beginning of his experiment. Entomologists used to think that in some way the silk helped the caterpillars to follow each other. This proved not to be so, although the caterpillars use the silk trail to return to their communal web. Because the caterpillars are blind and are said to lack a sense of smell, it is most likely they maintain their procession by touch. If a caterpillar is pushed out of line, the procession stops until that individual is back in its place, and then all set off again. It seems there is no question of there being an established leader. It is quite accidental which caterpillar starts the procession, and any one of the column can act as a leader.

The habit of the processionary moth caterpillar may seem odd. Yet the follow-the-leader behavior must have advantages for the species. For one thing, by using the silk trails they lay down and by following each other, these caterpillars are sure of reaching their feeding grounds and of returning to the security of their communal web, whether after feeding or to escape an oncoming storm, to the approach of which these caterpillars are seemingly sensitive.

Dart hairs

At first glance, it seems that the processionary behavior of these caterpillars must leave them vulnerable to predation. However, the larvae can produce slender irritating, or urticarial, hairs as a means of protection. These hairs are released into the air and can easily penetrate the skin surface of many animals, where they produce very severe irritation. Similarly these hairs may break off when a caterpillar is touched.

PRONGHORN ANTELOPE

THE PRONGHORN IS THE sole living representative of an ancient family, Antilocapridae, that arose and developed in North America. Usually called the pronghorn antelope, the pronghorn is not a true antelope, as these are found only in Eurasia and Africa. It stands 32–41 inches (81–104 cm) at the shoulder and is 3⅓–5¼ feet (1–1.6 m) in length, with a tail that grows 3–7 inches (7.5–18 cm) long. The pronghorn's weight is in the range of 79–154 pounds (36–70 kg), with the bucks (males) about 30 pounds (13.5 kg) heavier than the does (females). The upper parts of the coat are reddish brown to tan with a black mane and the underparts and rump are white, with two white bands across the neck.

Pronghorns have long, pointed ears and large eyes that are set out on the sides of the head, allowing a very wide range of vision. The buck has a black face and a patch of black hair on the side of the neck, characteristics that are less pronounced, or missing, in the female. Both sexes carry horns, which in the male are longer than the ears, consisting of a permanent, laterally flattened bony core, as in true antelopes, covered by a sheath of fused hairs. Like the antlers of deer, these horns are shed annually at the end of the breeding season, the bucks losing theirs first and the does shedding theirs a little later. The horns are erect, backward-curving and reach up to 20 inches (50 cm) in length, although most are about 15 inches (38 cm) long on average. The pronghorn's common name is derived from the short, forward-pointing branch, which is really part of the sheath, arising from the upper part of its horns.

Pronghorns live in rocky desert and grassland in western Canada, the western United States and northern Mexico.

Renowned for its speed

The pronghorn can leap up to 25 feet (7.6 m) at one bound. It is the swiftest mammal in North and South America, capable of cruising at a speed of 30 miles per hour (48 km/h), and is able to attain a speed of 50 miles per hour (80 km/h) over distances of up to ¾ mile (1.2 km). It is also

In winter the pronghorn antelope uses its forefeet to dig through snow cover to reach shrubs and also to scrape holes in which to deposit its droppings.

The black face mask and patches beneath the ears of male pronghorns are used in courtship and dominance displays.

PRONGHORN ANTELOPE

CLASS	**Mammalia**
ORDER	**Artiodactyla**
FAMILY	**Antilocapridae**
GENUS AND SPECIES	***Antilocapra americana***

ALTERNATIVE NAMES
Pronghorned antelope; American antelope

WEIGHT
79–154 lb. (36–70 kg)

LENGTH
Head and body: 3⅓–5¼ ft. (1–1.6 m); shoulder height: 32–41 in. (0.81–1.04 m); tail: 3–7 in. (7.5–18 cm)

DISTINCTIVE FEATURES
Coarse outer coat, areas of which can be isolated to protect skin from oncoming breeze in cold weather; reddish or tawny upperparts; 1 or 2 broad white rings around neck; white underparts; black cheek patches (male); laterally flattened horns (both sexes)

DIET
Cacti, lichens, sedges, grasses and shrubs

BREEDING
Age at first breeding: 15 months (female), 2–3 years (male); breeding season: July–October; number of young: 1 or 2; gestation period: about 250 days; breeding interval: 1 year

LIFE SPAN
Up to 11 years in captivity

HABITAT
Grassland, prairie and rocky desert

DISTRIBUTION
Western Canada south to northern Mexico

STATUS
Conservation dependent; estimated population: 1 million

Pronghorn antelope

a powerful swimmer. The woolly undercoat is covered by long, coarse guard hairs, which can be maintained at different angles by the flexing of certain skin muscles. Cold air is excluded when the hairs lie smooth and flat, but these can be raised to allow air movements to cool the skin in the heat of the desert sun. Cartilaginous pads on the hooves, particularly those of the forefeet, act like foam rubber soles, helping the pronghorn to travel quietly and quickly.

Winter nomads

The pronghorn eats a variety of low-growing grasses, shrubs, cacti and weeds and can get all the moisture it needs from this diet if necessary, although it will drink freely when water is available. The daily feeding range may be as much as 2 square miles (5.2 sq km). Apart from the old bucks, which are sometimes solitary, pronghorns are gregarious animals and roam in small, scattered bands throughout the summer. In winter they mass in herds of up to 100 or more and several times a year they shift from one area to another in search of food.

Solitary birth

The rut begins in late summer when fights break out between the bucks. When the harems, which may consist of up to 15 does, have been collected, mating takes place, the season lasting 2–3 weeks. For a first birth there is normally a single fawn, but in later births twins or, more rarely, triplets may be born. The doe seeks solitude for the birth,

in rocky areas or open country with low vegetation. Young pronghorns are born with a wavy grayish coat. They can walk within a few hours of birth and start grazing after 3 weeks. The coloring and texture of the coat breaks up the light, making the fawn difficult to distinguish and affording it some protection against predators. By the age of 3 months the first adultlike coat has grown and at 15–16 months the does mate, although the bucks probably do not breed until they are 2 or 3 years old. Pronghorns have lived up to 11 years in captivity.

White for danger

The pronghorns' main predator is the coyote, especially in winter, when the snow makes rapid movement impossible, and bobcats sometimes take the young. When a pronghorn becomes aware of danger, the hairs of its white rump patch are raised, alerting other pronghorns. This white flash can be seen by humans over 2 miles (3.2 km) away. Pronghorns have excellent distance vision. They are also naturally curious animals and will often approach an unfamiliar object if not startled by a sudden movement or alarmed by its scent. As is the case in many herbivores, the eyes of pronghorns are sensitive to movement. They will approach stationary predators unless alerted by some movement by the latter.

Curious pronghorn

The family Antilocapridae dates back 20 million years to the Middle Miocene period in North America, when the pronghorn population is estimated to have been 40 million. Hunting for sport, trophies and meat greatly reduced their numbers, which fell to 20,000 in 1925. Today, however, as the result of a policy of careful conservation, pronghorns are on the increase, the present population standing at about 1 million.

The pronghorn's acute inquisitiveness was quickly noticed by the early pioneers in North America. A pronghorn will inspect any moving object, such as a bush waving in the wind or small dust devils raised by the wind, or anything unfamiliar such as dogs, goats, cattle and even machinery. Early settlers attracted the pronghorn's attention by pushing a stick into the ground and tying a white handkerchief on it, which flapped in the wind. Its interest piqued, the pronghorn would come closer to inspect the moving object, bringing it within range of the settlers' guns.

Their tan and white coloration enables pronghorns to blend in well with their native prairie and grassland. The coat is unique among large North American mammals.

PRZEWALSKI'S HORSE

It is thought that Przewalski's horse once ranged across the entire Eurasian steppe, living in regions of scrubby grass and semidesert. By the 1970s it was extinct in the wild.

THE LAST SURVIVING wild horse, Przewalski's horse is a member of the subgenus *Equus*, a group that includes only Przewalski's horse (*E. przewalskii*), the tarpan (the extinct European wild horse, sometimes called *E. ferus*) and the domestic horse (*E. caballus*). Przewalski's horse is wild in the sense that is has never been domesticated, unlike the mustangs, *E. caballus*, of North America. Like the zebras of the subgenus *Hippotigris*, it is tamable only in exceptional cases.

Przewalski's horse is regarded as a subspecies by some authorities, named *E. caballus przewalskii* or *E. ferus przewalskii*. This is because there is no apparent barrier to hybridization with the domestic horse. Opponents of this view point out that Przewalski's horse has 66 chromosomes, as opposed to the domestic horse's 64. Hybrids, although they are fertile, possess 65 chromosomes; their offspring revert to 64 chromosomes and bear little resemblance to Przewalski's horse.

Przewalski's horse is distinguished by a dark stripe down the middle of its back. It has short legs and a robust build, with a heavy skull and a thickly shaped jaw. Its stiff mane lacks a forelock, and its tail extends from the body before broadening into a plume.

The story of Przewalski's horse is a remarkable tale of a species being brought back from the brink of extinction. The species would have disappeared forever had it not been for ambitious captive breeding and reintroduction programs.

Early decline

The history of Przewalski's horse, like its relationship to other horse species, is a little vague. When the species was discovered by Europeans in 1878, it was already in terminal decline, and surviving animals were limited to small enclaves in the remotest parts of Central Asia. It is thought that Przewalski's horse once ranged across the entire steppe belt of Eurasia, from China to the Ukraine. Fossil evidence supports this, but there are a few historical accounts of "wild horses" in Central Asia, and some cave paintings that are eerily suggestive of the horse's characteristic profile. Before scientists could learn much more about Przewalski's horse in the wild, the species continued to decline, and it was eventually confirmed extinct in the wild by the 1970s.

Ironically, one reason Przewalski's horse survives today is the rush of European collectors who rounded up many of the horses soon after

PRZEWALSKI'S HORSE

CLASS	**Mammalia**
ORDER	**Perissodactyla**
FAMILY	**Equidae**
GENUS AND SPECIES	***Equus przewalskii***

ALTERNATIVE NAMES
Przewalski's wild horse; Asiatic wild horse

WEIGHT
Average 550 lb. (250 kg)

LENGTH
Head and body: about 7 ft. (2.15 m);
shoulder height: about 4½ ft. (1.4 m)

DISTINCTIVE FEATURES
Low-slung, robust build; short, sturdy legs;
short, stiff mane without forelock; tan in
color with dark stripe down center of back

DIET
Coarse grasses and other small plants; bulbs

BREEDING
Age at first breeding: usually 2–3 years;
breeding season: spring; number of
young: 1; gestation period: 310–330 days

LIFE SPAN
Up to 35 years or more

DISTRIBUTION
Hustain Nuruu and Gobi National Parks,
Mongolia; Dzungaria Basin, northern China

HABITAT
Temperate semidesert and steppe

STATUS
Endangered; estimated population: 1,500

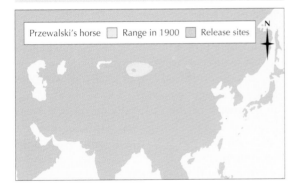

Przewalski's horse ☐ Range in 1900 ☐ Release sites

collectors, but also shot for food. The animals
sought refuge in remote, marginal grazing lands,
but were gradually excluded even from these by
ever increasing pressure from domestic sheep,
goats and horses. Their last refuges, in the Gobi
Desert, the Altai Mountains and the Dzungaria
Basin (Junggar Pendi, northwestern China), were
areas of cold semidesert and steppe, more or less
covered in hardy grasses. These are harsh places
with an extreme climate. While the summers are
quite warm, winter temperatures can plunge to
-50° F (-30° C). By the middle of the 20th century,
the wild population was confined to an area of
the Altai Mountains near the border of western
Mongolia and China, where the last confirmed
sighting occurred in 1968.

Meanwhile, even the captive stocks were in
peril. They fell to an all-time low of 13 animals,
including 3 stallions and 9 breeding mares, at the
end of World War II. It is from these individuals,
and one additional wild mare, that all of today's
1,500 or so horses are descended.

Worldwide breeding program

Such an extreme population bottleneck is often a
fatal problem for a wild population, and usually
leads to extinction. Thirteen animals represent
such a small gene pool that there is likely to be
insufficient genetic diversity to allow the species
to recover. However, Przewalski's horse largely
existed in captivity, and with the intervention of
humans and the management of a worldwide
breeding program, a healthy, out-bred population was maintained. The first step was to
compile a comprehensive stud book, work first

*There are about 1,500
captive and wild
Przewalski's horses
alive today. They are
all descended from just
14 animals following
an intensive captive
breeding program.*

the species was discovered. They shipped the
horses back to Europe, and in so doing, unwit-
tingly founded the captive breeding populations
that formed the basis of the species' reintroduction to the wild during the 1990s. During the 20th
century, wild populations were not only taken by

A female Przewalski's horse and her foal. Reintroduced horses have quickly adapted to the wild, forming herds and even demonstrating defense behavior in the face of wild predators such as gray wolves.

Back to its native home

The FPPPH eventually achieved its long-term goal of reintroducing Przewalski's horse back into its native Mongolian steppe habitat with the close cooperation of the Mongolian Association of Conservation of Nature and Environment (MACNE). A rare area of pristine steppe habitat was chosen, located near Mongolia's capital, Ulaan Baator. This area, Hustain Nuruu, was declared first a reserve, then a national park. Hunters and herdsmen were excluded from the area so that the Przewalski's horses would not have to compete for valuable grazing and watering places with domestic cattle. The herdsmen were partly compensated by their employment as wardens.

In a series of introductions during the 1990s, about 50 horses were flown into the Hustain Nuruu National Park and released, in harem groups. To ease the acclimatization process, the horses were kept safely in reserves when they first arrived. Experts estimate that the Hustain Nuruu site can sustain a population of 500 horses, and this population size is the long-term aim. With parallel projects underway in the Gobi National Park and in the Dzungaria Basin, the news is encouraging for Przewalski's horse at last.

undertaken by the Zoological Gardens in Prague, in 1960. These detailed records were used to choose mares and stallions with as little common ancestry as possible. If these animals were mated, the maximum possible genetic mixing would occur, so avoiding inbreeding depression. It was necessary to transport animals between the zoos of Western Europe, Russia, the Ukraine and North America. The Foundation for the Preservation and Protection of Przewalski's Horse (FPPPH) of the Netherlands encouraged and supported this exchange of animals, starting in the late 1970s.

Social behavior not forgotten

During the 1980s, groups of Przewalski's horses were released into reserves in the Ukraine, the Netherlands and Germany, as a step toward reintroducing the species to the wild. After several uninterrupted generations of captivity, the horses needed some experience of finding their own food and functioning naturally in their large, complex social groups. Remarkably, their social behavior had not been forgotten in their long period of captivity. It was found that Przewalski's horses form herds amounting to a stallion and his harem, typically four to six mares, and their foals.

Mating occurs in spring, shortly after the birth of the foals, so pregnancy lasts until the following spring. The mothers care for their foals for about 12 months. The young horses mature after 2 or 3 years, and then begin to leave the herd. Male juveniles strike out on their own, sometimes joining a bachelor group. Females leave to find a mate, perhaps a member of one of the bachelor herds.

Old habits die hard

Watching what the introduced horses do provides scientists with their first glimpse of what this species' behavior might be in its natural habitat. Exciting and tense moments such as the first herd's first encounter with a pack of gray wolves, *Canis lupus*, demonstrated that even predator defense behavior, something that these horses have not needed for 12 or 13 generations, has not been forgotten. The females immediately formed a protective ring around the foals while the stallion advanced aggressively toward the wolves, chasing them away.

Members of the first generation of horses born in Mongolia have now started breaking away from their natal herds, a natural behavior that limits inbreeding. There is a risk that these solitary roaming horses might come across herds of domestic horses and find a mate here, one of the processes that led to the original extinction of the species. This has already happened, but so far the horses are intensively monitored, and errant horses are removed and taken back to the heart of the National Park, where they can find Przewalski's horse mates.

PTARMIGAN

THE ROCK PTARMIGAN, *Lagopus mutus*, is 13–14 inches (34–36 cm) in length and has three seasonal plumages. In spring and summer the male has mottled gray upperparts, white wings and a black tail; the female is browner with black markings. In autumn both are grayer, but the female is lighter than the male. In winter both sexes are white all over except for their black tails, and the male has black eye patches. Young ptarmigan have the parents' summer plumage, except for dark primary flight feathers. The red comb (fleshy swelling) over the eye is larger in the male.

Rock ptarmigan range across much of northern Scandinavia, Siberia, Alaska, Arctic Canada, Greenland and Iceland, with isolated populations in the Scottish Highlands, the Pyrenees Mountains between France and Spain, the European Alps, Japan and Newfoundland.

The willow ptarmigan, *L. lagopus*, is a related form with a slightly more southerly range, in Eurasia and North America. The white-tailed ptarmigan, *L. leucurus*, is found in the Rocky Mountains, from Alaska to New Mexico. It is very similar to the rock ptarmigan, often living alongside it, but it is a distinct species.

Life among the rocks

Rock ptarmigan live among rocks with scanty vegetation at 2,000 feet (600 m) or more, in much more severe habitats than other members of the grouse family, Tetraonidae. When flying, they rise over humps and dip into hollows and can shoot up or down a precipice with equal ease. They walk or run with a rounded back and their tail down, often with a rolling gait. Ptarmigan crouch when alarmed, their plumage blending with the ground, and they only fly away when

Rock ptarmigan chicks are watched over by the female until they are 10–15 days old, when they can fly. If threatened by a fox, the female will feign injury to distract it.

The rock ptarmigan's white winter plumage is an adaptation to life in very cold climates. It also has thickly feathered toes that act as snowshoes.

ROCK PTARMIGAN

CLASS	**Aves**
ORDER	**Galliformes**
FAMILY	**Tetraonidae**
GENUS AND SPECIES	***Lagopus mutus***

WEIGHT
Male 14–21 oz. (400–600 g); female 12–16 oz. (350–450 g)

LENGTH
Head to tail: 13–14 in. (34–36 cm)

DISTINCTIVE FEATURES
Plump body; chickenlike bill; feathered toes; red comb (fleshy swelling) over eye. Male (breeding): gray with white wings and black tail. Female (breeding): browner overall. Male (winter): entirely white except for black eye patches and tail. Female (winter): lacks eye patches.

DIET
Buds, shoots, leaves, seeds and berries

BREEDING
Age at first breeding: 1 year; breeding season: eggs laid May–June; number of eggs: 5 to 8; incubation period: 21–23 days; fledging period: about 10–15 days; breeding interval: 1 year

LIFE SPAN
Up to 4 years

HABITAT
Tundra and mountain landscapes, including scree slopes and dwarf heath vegetation

DISTRIBUTION
Arctic and subarctic latitudes, including North America, Greenland, northern Europe and northern Asia; isolated populations in Alps, Pyrenees Mountains and Japan

STATUS
Scarce to locally common

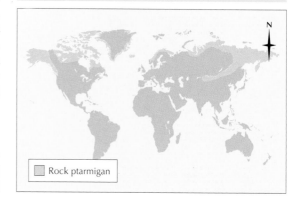

Rock ptarmigan

danger is imminent. They sunbathe and dust-bathe like other game birds, and roost in loose flocks. Family parties keep together until the autumn, when several may come together to form fairly large flocks. In winter ptarmigan burrow in snow. The calls, a cough or croak, a cackling snore and, when uneasy, a call like a clock being wound, are distinctly unmusical. The female may even hiss at an intruder.

In summer rock ptarmigan feed mainly in the early morning and in the evening, on the shoots, leaves and seeds of saxifrage, chickweed, whitlow grass, mountain avens and mountain sorrel. Berries eaten include bilberry, bearberry, cranberry and similar fruits, and a certain amount of insects are also eaten. The slightly larger willow ptarmigan has a more robust bill and eats a greater proportion of seeds.

Female raises the chicks

In spring the males display to the females by springing into the air with a sort of belching call and then dropping to the ground on outspread wings. They also chase the females with wings drooping and tail raised. The female makes the shallow scrape that serves as a nest among the short grass or low plants. At best, the nest is lined with a few pieces of grass, and five to eight eggs are usually laid at the end of May or in June, at 24- to 48-hour intervals. The eggs are pale yellowish white with dark red or chocolate blotches. When the clutch is complete, the female alone incubates for 21–23 days, the male keeping watch nearby. The chicks, brooded by the female and guarded by the male, can fly when 10–15 days old, becoming independent at 10–12 weeks.

Distraction displays

Ptarmigan are probably preyed on by birds of prey living in the same region. Ground predators include Arctic foxes, *Alopex lagopus*. If she has eggs or chicks, the female ptarmigan performs a distraction display (feigns injury). She crawls along the ground with tail fanned and wings thrashing to distract the fox. When the young ptarmigan are off the nest, she runs rapidly, flicking her wings to draw the fox away and then rising into the air when it nears her.

Adaptations to life in the cold

All three species of ptarmigan are clearly adapted to life in very cold climates. The whiteness of the ptarmigans' winter plumage does more than serve as camouflage. It also lowers the amount of heat lost from the birds' bodies. Ptarmigan also have heavily feathered toes that act as snowshoes. In particularly severe weather, ptarmigan use their strong legs and feet to burrow into the snow, where they are protected from the worst of the driving snow.

Ice Age relics

The rock ptarmigan's unusual distribution, all around the Arctic with a few scattered populations on high mountains such as the Alps farther south, is like that of a relict species. The presence of rock ptarmigan on the tops of the Pyrenees, Alps and other isolated mountainous areas suggests that the species was widespread over the northern half of the Northern Hemisphere during an Ice Age and that the birds moved northward and up into the mountains as the ice cap retreated in the interglacial period, when the warmer climate spread north.

A willow ptarmigan in intermediate plumage, between its summer and winter plumages. White feathers are limited to the legs, wings and and breast.

PUFF ADDER

A puff adder in the Kalahari Desert, southern Africa. The name puff adder is sometimes used for all species of the genus Bitis, *but is normally reserved for* B. arietans, *pictured above.*

THERE ARE 13 SPECIES of adders in Africa which belong to the genus *Bitis*. The English name puff adder has in the past been used for all members of this genus, but is more usually reserved for a single species, *Bitis arietans*. Adders of this genus range in size from the Gaboon viper, *B. gabonica*, which can grow to a total length of 6 feet (1.8 m), to the Namaqua dwarf adder, *B. schneideri*, which rarely exceeds 11 inches (28 cm). Puff adders themselves grow to about 4 feet (1.2 m) in total length. As well as being found across almost all of Africa, puff adders also occur in parts of Arabia. All of the *Bitis* species are stout-bodied snakes with short tails. They have a pit above each nostril, and it is possible that this organ can detect heat. The head is very broad compared with the neck, and is covered with small, overlapping scales. Some of these adders, such as the rhinoceros viper, *B. nasicornis*, found in West and parts of central Africa, have one or more erectile scales on the snout, which form horns.

The puff adder is pale brown or yellow brown with black, pale-edged V-shapes on the body. Gaboon vipers are even more striking. The head is buff-colored, and has stripes that run from the eye to the upper lips. The back has an attractive geometric pattern in which browns, cream and purple are the predominant colors. Some of the smaller species are more sober, usually having a background brown color with variable dark brown or black markings. Some of them have very restricted ranges: Péringuey's adder, *B. peringueyi*, and the Namaqua dwarf

adder, for example, are found only along a thin strip of coast in Namibia and South Africa, and the desert mountain adder, *B. xerophaga*, is found only in parts of the Orange River Basin in South Africa.

Melting into the background

Those species that live in desert and scrubland environments tend to have dull coloring, harmonizing with the soils on which they are living. So also do the Gaboon and rhinoceros vipers, in spite of their bright colors, for their color patterns are disruptive. The rhinoceros viper is even more brilliantly colored than the Gaboon viper, described above, with more purple, and blue as well, along with green triangles margined with black and blue on its sides. Both snakes are virtually invisible on the carpet of dead and green leaves on the forest floor. The smaller species live on sandy soils. Several of these are able to climb into bushes, but generally *Bitis* adders keep to the ground, hunting mainly during the night.

Inoffensive yet deadly

The broad head of the puff adder houses large venom glands, and, although the effect of this snake's bite is less rapid than that of a mamba or a cobra, it is just as deadly. Fortunately, puff adders strike only at desirable prey or in self-defense, and in this last case they need a fair amount of provocation. If their venom is slow-acting, it is nonetheless potent. Puff adders can give out as much as 15 drops of venom at a time, with 4 drops being enough to kill a human. However, snakes usually give a first warning by hissing. The hissing sound is produced by forcing air from the lungs and windpipe through the glottis, and puff adders are especially loud in this respect.

Beckoning their food

The food of these adders varies widely across the species. Small prey, such as a frog, is grabbed and swallowed without being poisoned. However, larger prey is struck with the fangs and allowed to run away to die. The snake later follows its trail to eat it. The snake drags the carcass into its mouth using the teeth in the lower jaw. Once part of the victim has reached the throat, muscular swallowing movements carry it down, the snake holding its head up to assist this. The fangs are large, especially in the Gaboon viper, where they

PUFF ADDER

CLASS	**Reptilia**
ORDER	**Squamata**
SUBORDER	**Serpentes**
FAMILY	**Viperidae**
GENUS AND SPECIES	***Bitis arietans***

LENGTH
Up to 4 ft. (1.2 m)

DISTINCTIVE FEATURES
Stout body; broad, rather flattened head; short tail; pale brown or yellow brown overall; black, pale-edged V-shapes on body

DIET
Rodents and other small mammals; also ground-living birds, lizards, frogs and toads

BREEDING
Breeding season: October–December; number of young: 70 to 80, sometimes more; hatching period: up to 1 year; breeding interval: often every 2–3 years

LIFE SPAN
More than 20 years in captivity

HABITAT
Most African habitats except very dry deserts and high mountains

DISTRIBUTION
Throughout sub-Saharan Africa; southern half of Arabian Peninsula

STATUS
Relatively common across much of range

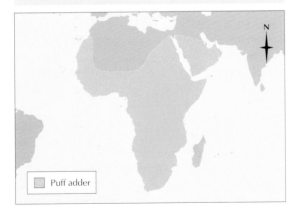

Puff adder

Puff adders are ovoviviparous, meaning that the female retains the eggs until they are ready, or nearly ready, to hatch. These snakes often produce very large numbers of young.

may be more than 2 inches (5 cm) in length in the biggest specimens. They serve to inject venom deeply into the body of the prey.

The puff adder and the Gaboon viper eat rats and mice, ground-living birds, frogs, toads and lizards. The Cape Mountain adder, *B. inornata*, feeds on the same but is known to also eat other snakes. The horned adder, *B. caudalis*, and the many-horned adder, *B. cornuta*, bury themselves in the sand, except for the eyes and snout, to catch lizards. The horned adder leaves the tip of its tail sticking out to attract victims.

A predator of many small animals, the puff adder has few adversaries itself. Those that do hunt these snakes include birds of prey, mongooses, warthogs and humans. Puff adders can store large quantities of fat, and this is sold by African herbalists as a cure for rheumatism.

Large numbers of young

Male puff adders engage in a combat ritual prior to mating. Puff adders are ovoviviparous, meaning than the eggs are hatched inside the mother so the young are born alive, or else they wriggle out of the egg capsule soon after it is laid. Mating is usually from October to December, the young being born in March and April. The young snakes are about 8 inches (20 cm) long at birth. Puff adders lay very large litters of young: 70 or 80 are not unusual, and a litter of 156 young recorded from a large female puff adder from East Africa is probably the record for any snake. Details of the breeding cycle vary from place to place. Many females breed only once every 2 or 3 years, and gestation can be up to a year.

Puff adders are relatively common in many places, and Gaboon vipers are common in some forest areas. Some of the smaller species have such restricted ranges that they are vulnerable, especially as a result of habitat fragmentation or destruction. Several of these species are protected in South Africa, as is the Gaboon viper because suitable forest habitats are now scarce.

PUFFBIRD

The spot-backed puffbird, Nystalus maculatus, *can be found in forest and arid scrubland in parts of Brazil, Bolivia and Argentina.*

THE PUFFBIRDS ARE NAMED after their chunky appearance and their loose, fluffy plumage. In Brazil they are known as *João bobo*, or silly John, in reference to their sluggish habits, for they are sometimes easy to approach. Puffbirds are related to the woodpeckers and jacamars, having two toes on each foot facing backward. The 33 species range from sparrow- to jay-sized. Their heads are large relative to the rest of the body, their wings are short and rounded, and their bills are stout.

Most puffbirds live in the Amazon Basin and northern South America, but a few live north of Panama. The white-necked puffbird, *Notharchus macrorhynchus*, ranges from Mexico to the Amazon and has striking black and white coloration. Most puffbirds are, however, brown or gray with streaks and speckles. The white-whiskered puffbird, *Malacoptila panamensis*, is 7½–8 inches (19–20 cm) long and ranges only as far south as Ecuador. The male is chestnut brown with tawny underparts, turning white on the belly, and a white patch on either side of the bill like a drooping mustache. The female is mainly grayish. The swallow-wing, *Chelidoptera tenebrosa*, has a build similar to that of these puffbirds but has long wings. It is like a stout swallow, about 6 inches (15 cm) long, and almost black, with a white rump and red belly. A small group of puffbirds are called nunlets, genus *Nonnula*, or nunbirds, genus *Monasa*, because of their dull brown and gray coloring, whose only relief is an orange bill.

Puffbirds occur in forests or forests edges at altitudes from sea level to 6,800 feet (2,000 m), living alone or in small groups. The nunbirds often travel in small flocks.

Tied to a perch

Puffbirds appear extremely lethargic, perching motionless with their feathers puffed out, but they are in fact alert to pounce on any insect that appears. The swallow-wings can usually be seen sitting in pairs on the bare branches of treetops, from which they launch graceful sorties against insects before returning to their perches. Puffbirds also snatch small animals from leaves and branches, and a few descend to the ground for their prey. On returning to the perch, the catch is pounded against the branch before being swallowed. Puffbirds' main food is large insects such as butterflies and locusts, spiders and sometimes lizards. They have been reported to catch lizards as large as themselves.

Puffbirds are usually silent, uttering no more than a few soft whistles and peeps. However, the white-fronted nunbird, *Monasa morphoeus*, makes a noisy whistle, sometimes uttered by 10 to 12 birds simultaneously. Puffbirds are unwary birds and readily perch in exposed places, refusing to fly far if disturbed. They often become considerably attached to their favored perches and use them year after year.

WHITE-NECKED PUFFBIRD

CLASS	**Aves**
ORDER	**Piciformes**
FAMILY	**Bucconidae**
GENUS AND SPECIES	***Notharchus macrorhynchus***

WEIGHT
About 4 oz. (110 g)

LENGTH
Head to tail: 9½–10 in. (24–25.5 cm)

DISTINCTIVE FEATURES
Heavy build with relatively large head; chunky bill; black crown; white forehead and collar; blackish upperparts; white underparts with broad black chest band

DIET
Mainly invertebrates; sometimes lizards

BREEDING
Age at first breeding: 1 year; breeding season: eggs laid March–June; number of eggs: 2 to 3; incubation period: not known; fledging period: not known; breeding interval: probably 1 year

LIFE SPAN
Not known

HABITAT
Forest edge, secondary growth woodland and plantations

DISTRIBUTION
Southeastern Mexico to northern Argentina

STATUS
Scarce to common

White-necked puffbird

Hidden entrances

Most burrow-nesting birds, such as kingfishers, jacamars and martins, make their burrows in the sides of steep banks where predators have difficulty in reaching them. In contrast, puffbirds usually burrow in gently-sloping ground, but effectively camouflage their nests. The white-whiskered puffbird hollows out burrows 18–22 inches (46–56 cm) long. The nest chamber is lined with leaves, and is camouflaged with a collar of dead leaves and sticks that blends with the surrounding leaf litter. The nest of the black nunbird, *Monasa atra*, is hidden beneath a large pile of sticks with a tunnel in the side that connects with the burrow.

Puffbirds eggs are incubated by both parents. The white-whiskered puffbird has an unusual pattern of incubation. The male sits from late afternoon to the following dawn, whereas the female sits only during the morning, the eggs being left unattended between these periods. The male broods the chicks for a few days while the female does all the feeding. When brooding has ceased, the chicks camouflage the nest further by sealing the entrance to the nest chamber each night with leaves from the nest lining.

Termite neighbors

Some species of puffbirds burrow in the nests of tree-dwelling termites. The male and female black-breasted puffbird, *Notharchus pectoralis*, take turns excavating a tunnel in a termite nest suspended on a bough. The tunnel widens into a nest chamber. The termites seal off their galleries from the puffbirds' tunnel so the two species leave each other alone.

The mustached puffbird, Malacoptila mystacalis, inhabits humid forests in the foothills of the Andes, in Colombia and Venezuela.

PUFFERFISH

A diver shadows an inflated spotted pufferfish, Arothron meleagris. This species can reach 19¾ inches (50 cm) in length.

PUFFERFISH, OR PUFFERS, are known by a variety of other names, such as balloonfish, swellfish, globefish and blowfish. All of these names make reference to the outstanding feature of pufferfish, that of being able to blow themselves up to twice their normal size or more. The related porcupine fish, covered elsewhere, also have this ability. When not inflated, a puffer-fish's body is flattened on the underside and rounded above. A pufferfish also has a large head and large, prominent eyes that give an air of perpetual surprise, or terror, depending on the observer's point of view. The dorsal fin is set far back, with the anal fin immediately below it, and the pufferfish uses these, together with the pectoral fins, in swimming. The skin is tough,

OCEANIC PUFFERFISH

CLASS	**Osteichthyes**
ORDER	**Tetraodontiformes**
FAMILY	**Tetraodontidae**
GENUS AND SPECIES	***Lagocephalus lagocephalus***

WEIGHT
Up to 6½ lb. (3 kg)

LENGTH
Up to 2 ft. (60 cm)

DISTINCTIVE FEATURES
Plump body, capable of inflation with air or water; large head; large eyes; small mouth; bright blue to steel blue back; pure white belly; small spines embedded in belly are erected when stomach inflated

DIET
Crustaceans and squid

BREEDING
Poorly known

LIFE SPAN
Not known

HABITAT
Tropical, subtropical and temperate seas and oceans; sometimes in estuaries

DISTRIBUTION
Atlantic, Pacific and Indian Oceans

STATUS
Common

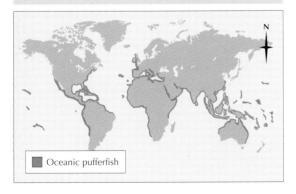

Oceanic pufferfish

A pair of freshwater pufferfish, Tetraodon palembangensis, *from West Africa.*

with small erectile spines that take the place of scales. Pufferfish that live in the open sea are often greenish or bluish black on the back. Those that live around coral reefs tend to be blue or purple brown with white stripes, spots and marblings. Generally the belly is lighter in color than the back. In some pufferfish the colors of the back extend to the belly with only a slight differ-ence in shade, whereas in others the belly is white. The largest species of pufferfish is the star pufferfish, *Arothron stellatus*, which grows to about 39 inches (1 m) in length.

Pufferfish live in tropical, subtropical and temperate seas, usually inshore among coral reefs, although several species are also found far from the coast. There are also freshwater puffer-fish, which occur well upriver in large rivers such as the Nile and Congo in Africa.

Unusual way of swimming
The shape of pufferfish indicates that they do not swim fast. The gill capacity is also low, showing that they use little energy when swimming. The swimming muscles normally found in the rear half of a fish have virtually disappeared, but those working the dorsal and anal fins are strongly developed. Pufferfish swim by waving the dorsal and the anal fins from side to side, the pectoral fins providing assistance, while the tail fin acts only as a rudder. These fish take prey that does not move or moves only slowly. They rely for protection on camouflage and inflating the body, although they probably have few predators besides sharks and barracudas, large specimens of which have been found with an inflated pufferfish impacted in the throat.

Inflated with water
A pufferfish inflates its body by distending the belly. When in water, an 8-inch (20-cm) puffer-fish takes in as much as a quart (about 1 l) of water, which passes into a sac leading from the

stomach and is kept there by muscular valves. When removed from the water, the threatened pufferfish responds by taking in air. Scientists do not know how long a pufferfish can maintain its defenses, but presumably a time must come when muscle fatigue causes the fish to get rid of the water it has taken in and return to normal size. It is also probable, as in the case of frogs and toads that inflate themselves as a protection against predators, that some time must elapse before the inflation reflex again becomes fully operative. This would be an extremely vulnerable period for a pufferfish.

Special crushing teeth

The teeth of pufferfish, like those of porcupine fish, are fused to form a beak. In both upper and lower jaws they form a cutting edge in front and a grinding plate behind. The pufferfish has has a cleft in the middle of the cutting portion of these dental plates, giving it the appearance of having four teeth, two in each jaw. This characteristic is the origin of the family name of Tetraodontidae, or "four teeth." The species living around coral reefs use these teeth to hack off pieces of coral. The coral is crushed and the living parts digested. Bivalve mollusks are likewise crushed, and pufferfish also feed on small animals such as crabs, sea snails and worms.

Spawning takes place in late spring and summer. It is not certain, even in those few species for which anything at all is known of the breeding, whether the eggs are laid on the bottom or are discharged and later sink. The eggs hatch in 4–5 days, and the baby pufferfish, ¼ inch (6 mm) long, can inflate their bodies almost from the time they are hatched.

Death by pufferfish

The spiny skin and inflatable body are a pufferfish's first lines of defense. Many pufferfish, though not all, also have a second line in the poison they carry. Known as tetrodotoxin, or TTX, it is one of the deadliest poisons in nature. Tetrodotoxin is also found in certain other animals, including the Californian newt, *Taricha torosa*, and the blue-ringed octopus, *Hapaloclaena maculosa*, which is found around Australia. Tetrodotoxin is a powerful nerve poison. In humans it blocks the passage of messages through the nervous system. Initial symptoms of TTX poisoning include tingling and numbness, and if enough of the toxin has been ingested, the result may be paralysis and possibly death. There is no antidote.

Not all parts of a pufferfish's body carry tetrodotoxin, which is usually located in the liver, reproductive organs and intestine, sometimes in the skin and occasionally in the muscles. In Japan pufferfish are eaten as a fashionable delicacy known as *fugu*. Specially trained chefs are employed to prepare them. Even so, many people die each year from tetrodotoxin poisoning after eating pufferfish.

A pufferfish inflates itself by distending the belly. It takes in water, which passes into a sac leading from the stomach and is kept there by powerful muscular valves.

PUFFIN

THE THREE SPECIES OF puffins are colorful auks, related to the murres or guillemots, the little auk or dovekie and the razorbill. The Atlantic puffin, *Fratercula arctica*, is 10¼–14 inches (26–36 cm) long, with a massive, brilliantly colored and decidedly bizarre bill. It is one of the most popular and well-known of all seabirds. The comical effect of its bill is enhanced by colored horny patches above and below the eyes. The Atlantic puffin's plumage is basically the same as that of other auks; black above and white underneath, with the black extending around the neck as a collar. The legs are bright orange and the sides of the face are white.

Bills molt in winter

The Atlantic puffin's triangular bill has red, yellow and blue stripes with thick yellow skin around the corners of the mouth. Outside the breeding season the basal part of the horny covering of the bill, including the blue parts and the yellow skin, are shed, leaving the base of the bill narrower and horn-colored. At the same time the red tip becomes yellow. The bill of the young Atlantic puffin is more conventional, narrower and plainly colored, the inner half grayish brown and the outer half reddish brown.

The Atlantic puffin breeds along the coasts of the North Atlantic from Greenland south to the Gulf of St. Lawrence in the west and from Spitsbergen and Novaya Zemlya south to the British Isles and northern France in the east. Some puffins spread as far south as the Canary Islands and into the Mediterranean as far as Italy. British puffins have been found wintering in American waters, but not all puffins migrate away from their breeding places. Atlantic puffins regularly spend the winter in Baffin Bay, and in mild winters they stay near Amsterdam Island, north of Spitsbergen, despite the low temperatures and continual darkness of the arctic winter.

Pacific puffins

The horned puffin, *F. corniculata*, which lives in the North Pacific and is a close relative of the Atlantic puffin, has fleshy growths over the eyes and differs in the coloring of the bill. It breeds on either side of the Bering Sea, in easternmost Siberia and Alaska. Another Pacific puffin is the tufted puffin, *F. cirrhata*, which in summer is all black except for a white face and long tufted feathers sprouting from above the eyes. Its bill is mainly red, with a greenish base. The tufted puffin is found mainly in the Bering, Okhotsk and Arctic Seas.

Feet serve as air brakes

From spring until the end of the breeding season puffins contribute to the masses of auks of several species that fly continually to and from their traditional nesting cliffs. Instead of shuffling on their haunches like other auks such as the murres (discussed elsewhere), puffins walk easily with a waddling gait.

When puffins take off from the cliffs, their wings appear to be too small to support them, and they plunge steeply until their rapidly whirring wings become effective. When they are cornering in flight or coming in to land, their large webbed feet are spread out to help in steering and braking.

An adult tufted puffin in summer plumage. Most tufted puffins are found in the North Pacific, but they also wander as far south as California outside the breeding season.

Fish stacked in bill

Puffins feed on small fish, such as sand eels and young cod, together with small squid, crustaceans and other planktonic animals. Outside the breeding season puffins go far out to sea, usually out of sight of land. Food is caught by diving, the puffins swimming underwater with their wings. If puffins have chicks to feed, they carry their catch back in the bill. Puffins are quite tame on their breeding cliffs and can be watched from close quarters landing with fish draped crossways in their bills.

Puffins can carry up to 30 fish in this way, but this is exceptional. How they arrange the fish in the bill is still a mystery. Presumably each fish is killed by a nip with the bill, but how it is then placed alongside ones caught previously without dropping them is difficult to visualize. The tongue and the serrated floor of the upper mandible (bill-half) may be used to maneuver and hold them. Paintings of puffins with fish neatly arranged in the bill head-to-head or alternating

An Atlantic puffin touches down at its nest with a full load of sand eels.

ATLANTIC PUFFIN

CLASS **Aves**

ORDER **Charadriiformes**

FAMILY **Alcidae**

GENUS AND SPECIES *Fratercula arctica*

ALTERNATIVE NAME
Common puffin

WEIGHT
Average 16¼ oz. (460 g)

LENGTH
Head to tail: 10¼–14 in. (26–36 cm); wingspan: 18½–24¾ in. (47–63 cm)

DISTINCTIVE FEATURES
Squat body; bull neck; massive, triangular, laterally compressed bill; short orange legs, set far back on body. Summer: bright red, yellow and blue bill with fleshy yellow skin around base; white face and belly; black upperparts and collar. Winter: bill loses horny sheaths, becoming narrower and paler.

DIET
Mainly fish; also squid and small crustaceans

BREEDING
Age at first breeding: 5 years; breeding season: eggs laid May–June; number of eggs: 1; incubation period: 40–43 days; fledging period: 42–56 days; breeding interval: 1 year

LIFE SPAN
Up to 25 years

HABITAT
Summer: coastal waters off rocky coasts and nearby islands; nests on grassy slopes at top of coastal cliffs. Winter: farther offshore.

DISTRIBUTION
North Atlantic and Arctic Oceans

STATUS
Common

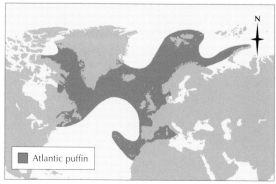

Atlantic puffin

head-to-tail are based on imagination. Working the fish into a pattern would be very difficult and would serve no useful purpose.

Declining numbers of sand eels, now being fished commercially in the North Atlantic, are depriving many Atlantic puffins of food with which to rear their young. This is having a serious effect on breeding success in some areas.

Nest in clifftop burrows

When puffins arrive at their breeding ground, they start digging burrows or clearing out old ones. They dig with their heavy bills and scrape the loosened soil out with their feet. In large colonies burrowing can be so extensive as to cause a landslide. Puffins sometimes take over shearwater or rabbit burrows.

Puffins arrive at the breeding grounds already paired, but there is a considerable amount of displaying around the burrows. The large colorful bills are used as signals, being thrust forward in threat or shaken in appeasement. Mating takes place on the water after the male has chased the female.

A single egg, white with faint markings, is incubated for 40–43 days. The parents share the task, but at intervals they leave the egg and parade together outside the burrow. The chick is fed on fish by both parents, but when about 6–8 weeks old it is deserted by them, and the adults go out to sea to molt. During this time the parents become flightless. The chick stays in the burrow for another week and then flutters down from the cliffs and paddles out to sea. The young puffins leave the burrows at night, when there is less danger from large gulls, jaegers and skuas. Until they can fly they avoid danger by diving. Seven or eight weeks is a very long fledging period for an auk, but puffins are reared in the safety of a burrow, whereas murres, razorbills and other auks breed out on cliff ledges where their chicks are vulnerable to predation. Those auks also leave the nest before they can fly but, unlike puffins, they are not independent and remain in the care of the adults.

Conservation

A certain number of puffins fall prey to gulls, jaegers and skuas. On some offshore islands, for instance, the cliffs may become littered with the remains of puffins eaten by great skuas, *Catharacta skua*. However, puffin numbers are more severely reduced when rats are introduced into their breeding grounds. At one time the Atlantic puffin population of Ailsa Craig in Scotland was described as phenomenal, but in 1889 rats came ashore from a wreck and the population has since declined almost to extinction.

Recently there has been another threat to puffins and other auks. Oil pollution is particularly serious to auks because of their sociable lifestyle and their habit of diving out of trouble and resurfacing in the oil. They are especially vulnerable when flightless during the molt.

A small gathering of Atlantic puffins near their breeding burrows. Puffins burrow several feet into the loose soil at the top of cliffs.

PUMA

The puma ranges from Canada southward to Patagonia, one of the largest distributions of any mammal. It varies greatly in both color and size, the largest animals being found farthest from the equator.

IT IS USUAL TO SAY THAT this animal is known in Britain as the puma and in the United States and Canada as the cougar. However, the puma has many common names, another of which is the mountain lion, which gives a clue to the appearance of this large cat. It looks like a lioness, its coat being of short, close fur, although this becomes longer and thicker in winter. It is uniformly tawny to silvery gray in color and some animals have banding on the upper parts of the legs. Melanism (all-black individuals) is also common. The maximum size recorded for a male was 9½ feet (3 m) long, of which 3 feet (90 cm) was tail, and 265 pounds (120 kg) in weight. However, there is much variation, from as little as 5 feet (1.5 m) in total length and 75 pounds (34 kg) in weight. Females are usually smaller than males. Pumas have the longest legs of any cat and broad paws for running in snow.

The puma ranges from western Canada to Patagonia at the southern tip of South America, and is found throughout most of the central United States. Individuals living farther from the equator tend to be larger than those living close to it. The puma lives in a variety of habitats including desert, prairie and montane forests, in addition to tropical forest and tundra. Occasionally it strays into the suburbs of towns or cities.

Although there is only one species of puma, 30 subspecies have been named, based on differences in size and color, a clear indication of how variable these two features are.

Powerful caterwauling killer

The puma is known for its remarkable strength and stamina. It is said to cover up to 20 feet (6 m) in one bound, and a leap of 40 feet (12 m) has been recorded. It can leap upward to a height of 15 feet (4.5 m) and has been known to drop to the ground from a height of 30 feet (9 m). Like many other members of the cat family, the puma leads a solitary life, keeping very much out of sight. A puma will often kill and drag its prey on the ground until it has reached cover. One puma was known to drag a carcass three times its own weight over the snow. A puma will travel 30–50 miles (48–80 km) when hunting. Its trail is marked by the remains of prey lightly buried in the ground and by the scratchings where it has scraped earth over its urine or dung.

PUMA

CLASS **Mammalia**

ORDER **Carnivora**

FAMILY **Felidae**

GENUS AND SPECIES ***Puma concolor***

ALTERNATIVE NAMES
Cougar; mountain lion; panther

WEIGHT
Male: 117–160 lb. (53–72 kg).
Female: 75–106 lb. (34–48 kg).

LENGTH
Male, head and body: 3⅗–6⅖ ft. (1–2 m); shoulder height: 1⅔–2⅔ ft. (50–80 cm). Female, head and body: 3⅕–5 ft. (1–1.5 m); shoulder height: 1⅔–2⅔ ft. (50–80 cm).

DISTINCTIVE FEATURES
Adult: resembles lioness; uniform tawny to silvery gray coat; all-black coat is common; upper legs sometimes banded; long legs with broad paws. Young: spotted coat; ringed tail.

DIET
Mammals such as deer and rodents; also birds, amphibians, insects and carrion

BREEDING
Age at first breeding: 2 years; breeding season: all year, most births in warm months; number of young: 2 to 5; gestation period: 87–95 days; breeding interval: 18 months

LIFE SPAN
Usually up to 12 years

HABITAT
Desert, prairie and montane forests; also tropical forest and tundra; sometimes suburbs

DISTRIBUTION
Western Canada, south through central U.S. to southern tip of South America

STATUS
Uncommon; endangered in North America

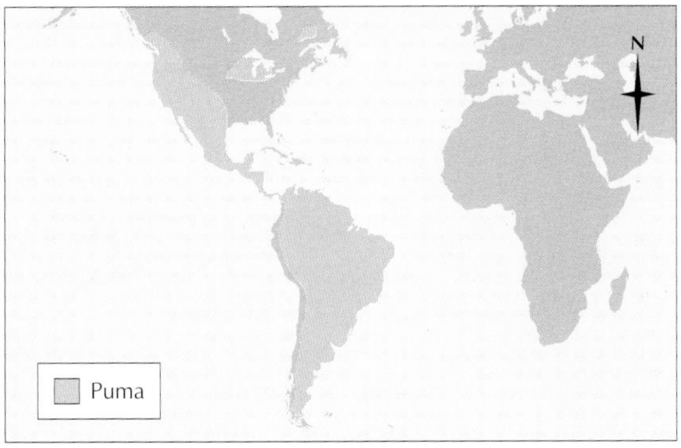

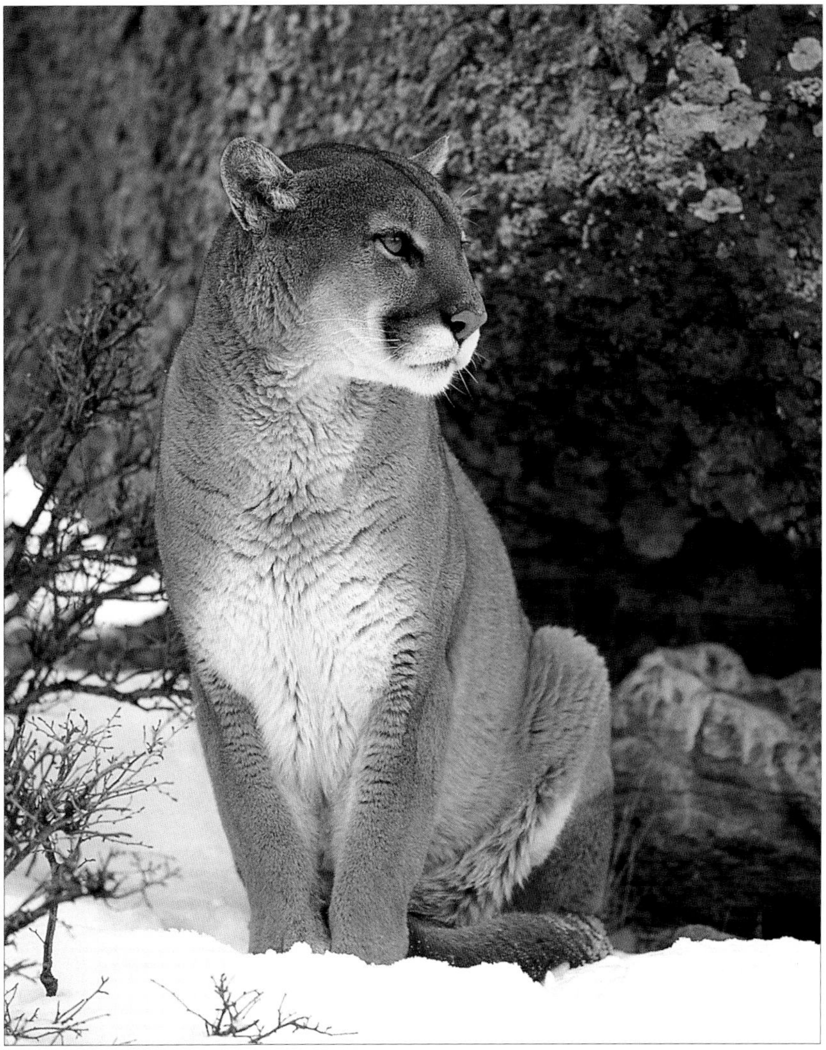

A puma's broad paws are good for traveling over snow in winter. The puma is a generalist predator hunting large prey such as deer in the north, but taking smaller prey in the Tropics.

Some of the puma's common names, such as catamount, refer to its voice, but there are differences of opinion as to how much this is used. Pumas cannot roar, but they are a lot more vocal than was once thought and have many different calls. They are said, for example, to emit bloodcurdling screams, especially during the breeding season, and there have been many vivid descriptions of this call. The evidence suggests that both male and female pumas scream, although not frequently. Like domestic cats, they normally purr when contented.

Another aspect of the puma's way of life that is much debated has to do with its attitude toward humans. It has been represented as highly dangerous, and as a generalist predator has been known to attack humans. However, most surveys show that attacks on humans are so rare as to be almost negligible. Between 1998 and 2000 there were nine such attacks. It seems likely that the puma's screaming, together with its habit of stalking people, may have contributed to an undeserved reputation for ferocity. The stalking seems to arise from an overwhelming curiosity, and many naturalists have told how

Although the puma is now protected across much of its range, the main threat to the species remains ranchers who may shoot these animals if they are seen to be taking livestock.

they have been followed by a puma for great distances, the cat melting into the undergrowth every time they look around. This habit is also common in other large cats.

Controlling the deer

The puma's main prey is deer, particularly in the northern part of its range, and these may make up 50–75 percent of its food. Wherever the puma is killed off, deer multiply rapidly. It also takes a variety of small mammals, including porcupines and rodents, along with insects, birds and amphibians. It sometimes attacks domestic stock, such as sheep and goats, as well as horses and cattle, and it will take carrion. Puma's stalk their larger prey, suddenly pouncing on their backs with a powerful leap, often with such violence that the animal is carried up to 20 feet (6 m) along the ground. The typical method of killing is by a bite into the throat.

Spotted kittens

Pumas breed year-round, although most births take place in the warmer seasons. After a gestation period of 87–95 days a litter of two to five cubs is born. They are blind at birth and have spotted fur and a ringed tail. The eyes open at 10–14 days and the cubs are weaned after 1–3 months. The period seems to vary, as does the

time they stay with the mother, which may be up to 2 years. As they mature, the cubs lose their spots and the rings on the tail. Pumas usually live 8–12 years, but very old animals of up to 18 years have been found.

Hunted by humans

Pumas have few predators apart from humans, but jaguars and pumas often fight where their ranges coincide. Wolverines will attack pumas, and grizzly bears are also credited with doing so. However, these hazards are very trivial compared with human activities, which have wiped out the puma or seriously reduced its numbers in most parts of its range. In the past pumas have been killed with traps and hunted, especially with dogs, which will usually tree the puma, thus putting it at the mercy of the hunter. In some places and at various times bounties have been paid, the main complaints against this large cat being made by ranchers because of its attacks on their domestic stock.

As a result, the puma is now endangered in North America and is protected across much of its range. Some subspecies, for example *Puma concolor coryi*, are critically endangered. Populations are very sparse, and the total number of individuals has been estimated at 500,000, although this is unlikely to be accurate.

PURPLE EMPEROR

A LARGE, SHOWY AND unfortunately rare butterfly, the purple emperor is named after the purple iridescence on the wings of the male. This effect can be seen only when the butterfly is viewed from a particular angle. Otherwise this insect is dark brown, almost black, with a line of white patches and an orange, black and purple eyespot on each hind wing. The white patches are occasionally absent or partially missing, and purple emperors with this latter patterning are known as the iole variety. The underside of the purple emperor's wing has an intricate pattern of brown and gray with bands of white. The female is very much like the male but lacks the iridescence and is slightly larger. Her wingspan is 3 inches (7.5 cm). The male's is 2½–2¾ inches (6–7 cm). Purple emperors occur throughout western Europe and temperate Asia as far as Japan.

Sapsuckers

The purple emperor's preference for deciduous woods is one factor that limits its distribution, because these woods become scarcer year by year as a result of human activities. The purple emperor does, however, survive in woodlands that have been stripped of all tall trees. It is on the wing in July and the first half of August, but unlike so many other butterflies, it is not attracted to flowers. Both males and females feed on the sap that oozes from wounded oaks. Males are also drawn to muddy tracks, road surfaces and animal excreta, from which they extract moisture and mineral salts. At most times of the day, but most frequently in the later morning, males can be found at ground level. However, they spend a lot of time in flight and generally head toward the highest point of the locality.

Sluglike caterpillars

Female purple emperors lay their eggs in the early stages of sunny afternoons on the upper surfaces of shaded leaves deep in the growth of host plants. For the purple emperor, these hosts are the goat willow (*Salix caprea*), gray sallow (*S. cinerea*) or crack willow (*S. fragilis*). Each female ranges over a considerable area, selecting egg-laying sites and depositing a single egg at each one. A female may revisit a bush and lay a second egg on a leaf already bearing one.

Spot the difference! This is not a purple emperor but a lesser purple emperor, Apatura ilia, a close relative. The two species are similar in appearance but the lesser has a red-ringed black spot on each forewing, which the purple emperor lacks.

PURPLE EMPEROR

PHYLUM	**Arthropoda**
CLASS	**Insecta**
ORDER	**Lepidoptera**
FAMILY	**Nymphalidae**
GENUS AND SPECIES	*Apatura iris*

LENGTH
Adult wingspan: 2½–3 in. (6–7.5 cm)

DISTINCTIVE FEATURES
Adult: undersides of both sexes shaded brown with white markings; upper sides blackish brown with white markings; in sun, male shows iridescent purple with orange, black and purple eyespot on each hind wing. Caterpillar: green body tapering toward end; paler head with 2 long, pinkish-tipped horns pointing forward.

DIET
Adult: both sexes attracted to sap from oak trees; male also to moisture and minerals from road surfaces, muddy tracks and animal excreta. Caterpillar: goat willow (*Salix caprea*), gray sallow (*S. cinerea*) and crack willow (*S. fragilis*).

BREEDING
Eggs laid singly; hatching period: about 9–10 days; winter spent as caterpillars; emergence of adults following July–August

LIFE SPAN
More than 1 year

HABITAT
Temperate deciduous forest and woodland

DISTRIBUTION
Western Europe east through temperate Asia to Japan; Asian range poorly known

STATUS
Generally uncommon; rare in many areas

Purple emperors can be elusive butterflies. Males (above) spend much of their time in the treetops, where they gather and soar, often around a single tree. Females visit this tree to find suitable mates before flying off to lay their eggs.

Other females may also use that leaf, so it is possible to find several eggs on one leaf. The egg is approximately 1 millimeter high and dome-shaped, with about 14 radially arranged ridges. At first the egg is bluish green, but then the base becomes dark purplish brown, and just before hatching it turns black. The caterpillar emerges 9–14 days after the egg has been laid, depending on environmental conditions. It is yellow with a black head and measures just a little over 2.5 millimeters in length.

After 10 days of eating the leaf on which the egg was laid, the caterpillar sheds its skin. It is now green, the same color as the leaf, and it has a pair of horns projecting from the head like a slug's tentacles and a body tapering to a point at the rear end. The caterpillar continues to feed throughout the summer and into autumn and changes to brown, so matching the autumn leaves. In October it retires to a twig or a fork between two twigs and spins a mat of fine silk on which it rests for the winter. The following April the caterpillar begins feeding again on the fresh leaves, growing to 1¾ inches (4.5 cm) before pupating in June. Pupation takes place on the underside of a leaf, where it lays a mat of silk, running more silk up the leaf stem to the twig, presumably to act as an anchor. The chrysalis hangs from the silk mat by a number of small hooks. Just before pupating, the caterpillar changes to a very pale green, matching the underside of the leaf and making the chrysalis very difficult to find. The adult butterfly emerges in 2–3 weeks, once again depending on environmental conditions.

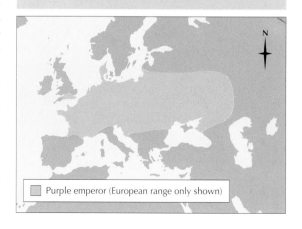

Purple emperor (European range only shown)

PURPLE SEA SNAIL

SOMETIMES CALLED THE violet sea snail or the floating shell of the high seas, the purple sea snail resembles a garden snail or a periwinkle. The body is purple and the shell, which is blue, lavender, violet or purple, is up to 2½ inches (6 cm) across and paper-thin, though not transparent. The mollusk lives almost permanently upside down in very large numbers in the Atlantic, Indian and Pacific Oceans, but sometimes drifts onto temperate shores.

The purple sea snail forms part of the oceanic plankton. It floats upside down at the surface on a bubble raft, which it makes by secreting, from its foot, a froth that traps bubbles of air. The tip of the foot is pushed out through the surface film. It closes around a bubble of air and brings it down to the middle of the foot where it is coated with mucus that hardens the wall of the bubble and glues it to others. The raft is usually three or four times the size of the snail.

Feed on jellyfish

Purple sea snails feed on other animals in the plankton, such as crustaceans, including barnacle larvae, and jellyfish, including members of the genus *Velella*, relatives of the Portuguese man-of-war. Velellas form colonies of polyps, with tentacles and stinging cells hanging down from a platelike float. While the purple sea snail feeds it squirts out a violet dye, which seems to anesthetize velellas and prevent their polyps and stinging tentacles from latching onto and harming the snail.

Purple sea snails also eat copepods, marine insects of the genus *Halobates* and other flattened disclike jellyfish of the genus *Porpita*. Purple sea snails are also known to be cannibalistic. They have no eyes but have a black tentacle near the mouth that may be an organ of smell by which the snails detect the presence of prey.

Purple sea snails lay their eggs on the under-surfaces of velella floats or bubble rafts. Most species lay eggs in rows of capsules, numbering 200 to 600 according to the species and the size of the individual laying the eggs. There may be as few as 17 eggs per capsule in one species and as many as 5,500 in another. The purple sea snail that most commonly reaches British shores gives birth to live young, or ciliated larvae, called veligers. Young purple sea snails are male at first, becoming female later in life.

When it is alive, a purple sea snail has its own raft of bubbles to keep it afloat upside down. These delicately colored, empty shells are dead specimens.

PURPLE SEA SNAILS

PHYLUM	**Mollusca**
CLASS	**Gastropoda**
ORDER	**Mesogastropoda**
FAMILY	**Janthinidae**
GENUS AND SPECIES	***Janthina janthina, J. exigua, J. globosa*** and ***J. pallida***

ALTERNATIVE NAMES
Violet snail; violet sea snail

LENGTH
Shell: up to 2½ in. (6 cm)

DISTINCTIVE FEATURES
Very fragile, coiled shell is purple, lavender blue or violet; floats beneath raft of bubbles

DIET
Other planktonic animals; goose barnacles

BREEDING
Sexes separate, but all snails start life as males, becoming female later. Number of eggs: 200 to 600, but larvae hatch directly from female in *J. janthina*.

LIFE SPAN
Not known

HABITAT
Floats at sea surface

DISTRIBUTION
Virtually worldwide in tropical seas

STATUS
Abundant

PURSE SPONGE

Purse sponges inhabit rocky crevices so that they do not dry out when the rocks to which they are anchored are exposed by the receding tide waters.

THE PURSE SPONGE is one of the few sponges, apart from the various kinds of bath sponges, to be given a common name. It is relatively small, being usually no more than 2 inches (5 cm) high, although very exceptionally it reaches a height of 6 inches (15 cm). However, perhaps more completely than any other species, the purse sponge illustrates the variable nature of sponges, especially with regard to their asexual reproduction.

The creamy white purse sponge often has a shape similar to that of a flattened vase. However, the sponge's outline often varies greatly from this simple general form, becoming regular and more folded on itself.

Purse sponges live between tidemarks, especially on rocky shores, from mid-tide level to low-water spring tides. Occasionally they live in inshore waters down to 60 feet (18 m). They extend around the coasts of Europe and into the Arctic, more sparingly in the Mediterranean, and in scattered localities in other parts of the world.

Probably feed on bacteria

Purse sponges live under stones, on bunches of seaweed, between rocks, in caves or in cavities formed by boulders, so that they never completely dry out, even when the shore is uncovered by the ebbing tide. Almost invariably they hang with their mouths down.

The outer surface of a purse sponge is perforated with microscopic pores, each leading to a thimble-shaped flagellated chamber lined with collared cells called choanocytes (see the separate sponge article). Water is drawn in through the pores by the combined beating of the whiplike flagella of the choanocytes and leaves by the mouth of the thimble into the central cavity of the sponge. From there the accumulated effluents from all the flagellated chambers spill out through the mouth. In its passage through the sponge the water gives up oxygen, and fine particles trapped inside the collars of the choanocytes are absorbed. How this is done and how food material is distributed throughout the body of the sponge remains a matter of scientific debate. There is uncertainty too over the purse sponge's feeding habits. Large purse sponges are found in estuaries and harbors where the water is heavily polluted with sewage and garbage. It is logical to assume, therefore, that the food of a purse sponge is composed of either very minute plants and animals or microscopic particles from dead plants and animals or bacteria. It is possible that it feeds on all of these substances. There is little doubt, however, that all sponges, like other particulate feeders such as mussels, are scavengers that help to purify the water in which they live.

The purse sponge has a skeleton of calcite spicules (slender, pointed projections). The spicules have three rays (triradiate) to which a fourth ray (quadriradiate) is sometimes added; some of the spicules are needle-shaped rods (monaxons). The spicules are arranged in a fairly constant pattern in the skin and around the flagellated chambers and may serve to support the tissues.

Sex without sex organs

Purse sponges are hermaphrodites, sperm and eggs being produced by each individual. There are no special reproductive organs, but in spring certain of the body cells become swollen with food granules and increase greatly in size. At the same time other, smaller cells migrate through the tissues and arrange themselves around each enlarged cell to form a capsule enclosing it.

Some of the enlarged cells become eggs, whereas others divide repeatedly to form a mass of small cells, which are the sperm. The latter are

PURSE SPONGE

PHYLUM	**Porifera**
CLASS	**Calcarea**
ORDER	**Sycethidae**
FAMILY	**Orantidae**
GENUS AND SPECIES	***Scypha compressa***

LENGTH
Usually up to 2 in. (5 cm); exceptionally up to 6 in. (15 cm)

DISTINCTIVE FEATURES
Coarse surface; flattened, vaselike shape with small hole at one end; off-white coloration; has numerous folds and lobes if reproducing asexually

DIET
Filter feeder on microscopic particles

BREEDING
Sexual reproduction: hermaphroditic, producing both sperm and eggs. Asexual reproduction: wide variety of methods, but mainly budding and fragmentation.

LIFE SPAN
Not known

HABITAT
Between tidemarks on rocky shores

DISTRIBUTION
Common in northern Europe; wider distribution unclear

STATUS
Locally common

Multiply by fragmentation

The smaller purse sponges have a simple outline, but in the majority the body is folded on itself in various ways or it has holes in it or several secondary vaselets around its margins.

These shapes represent the various ways the sponge reproduces asexually. For example, at one point on the surface of the sponge a lobe grows out, which folds over onto the body. A line of weakness develops along the fold, and in due course the sponge splits along it. The lobe then drops away and falls to the bottom. The split on the parent body heals, as does the split on the fragmented piece, which at the same time grows a stalk that becomes fastened to a suitable rock surface.

In another method of reproduction, the vaselets grow on the margin and then drop off. The broken surfaces heal and each vaselet grows a new stalk. There are eight or nine different methods by which a purse sponge can throw off bits of itself, each fragment then developing into a new purse sponge.

In this photograph several exposed clusters of creamy white purse sponges are attached to strands of seaweed.

later liberated by the breakdown of the capsule. They are carried into the surrounding water by the outgoing current and swim to another purse sponge, entering it on the ingoing current and fertilizing the eggs in it. The fertilized egg subdivides repeatedly to form an egg-shaped mass of very small cells. Those cells at the smaller end throw out whiplike flagella. These flagella begin to beat, and the larva rotates and then bursts from its capsule and swims out to the exterior. Each larva swims about for less than a day, after which time the flagella are withdrawn. Without the flagella the larva sinks to the bottom and settles on a rock or a piece of seaweed, to start life as a new purse sponge.

The only known disease from which living sponges suffer is a fungus. They have only a few predators, including nudibranches and winkles.

PUSS MOTH

Camouflage is an important protective mechanism for the adult puss moth.

Closely related to puss moths are kitten moths, genus *Harpyia*, which are smaller and have darker bodies and more heavily patterned forewings.

Caterpillars look spectacular

Like all the prominent moths of the family Notodontidae, the puss moth is a night flier. It is readily attracted to light and can be found in May and June, in places where poplars, sallows and willows grow. The reddish brown hemispherical eggs are laid on the leaves of these trees. They are placed, usually in pairs, on the upper surfaces of the leaves.

When small, the caterpillars are black with a pair of long tails. Some zoologists claim that these tails are an alternative source of the puss moth's name. As they grow, the caterpillars develop a pattern of green and brown, and when fully grown, they are 2–2½ inches (5–6.5 cm) long. While feeding, the body appears hunchbacked. Four pairs of fleshy legs known as prolegs support the rear part of the body, and from either side of the tip of the abdomen rise the two long tails, broad near the base, with whiplike tips. Along the head and back runs a broad band of purplish brown edged with white, which forms a saddle in the middle of the body. This pattern is an example of disruptive coloration, breaking up the outline of the caterpillar so that it is difficult to see among the leaves and is easily mistaken for a withered leaf.

The caterpillars feed throughout July and August and sometimes into September. Then they select crevices in the bark of the trunk or branches of the tree in which they were feeding. As the caterpillar weaves its cocoon of silk, it incorporates scraps of bark and chewed wood, making the cocoon barely distinguishable from its background.

Cutting its way out

The cocoon is a tough structure, and the caterpillar leaves the skin at the head end thinner than the rest so that the adult moth can escape next summer. To assist its emergence, the moth has two mechanisms for opening the cocoon, one mechanical and one chemical. When it bursts out of the pupa case or chrysalis within the cocoon,

THE PUSS MOTH BELONGS TO a family of species known commonly as prominent moths. These have a coating of fluffy hairs similar in texture to those of a cat. The family is named after the prominent tuft of scales on the rear margin of the hind wings. The wings and body are whitish, the thorax has dark spots and the abdomen has broad dark transverse bands. The wings have yellowish veins with black branches and the forewings have patterns of wavy lines. Puss moths, *Cerura vinula*, are about 1⅓ inches (3.5 cm) from head to the tip of the abdomen. The larger females have darker hind wings. Puss moths are found in most of Europe, east across Asia to Japan, and in North Africa.

PUSS MOTH

PHYLUM	**Arthropoda**
CLASS	**Insecta**
ORDER	**Lepidoptera**
FAMILY	**Notodontidae**
GENUS AND SPECIES	***Cerura vinula***

LENGTH
Adult (moth). Head and body: 1⅓ in. (3.5 cm); wingspan: about 2¾ in. (7 cm). Larva (caterpillar). Head and body: up to 2–2½ in. (5–6.5 cm).

DISTINCTIVE FEATURES
Adult: fluffy hairs on body; whitish overall with dark spots on thorax and broad dark bands on abdomen. Larva: very large size; bright green overall; red "face" with 2 black "eyespots;" purplish brown band with white borders runs along center of head and back; 2 pairs of fleshy legs; 1 pair of long, whiplike tails at tip of abdomen.

DIET
Larva: leaves and shoots of sallow, willow and poplar trees.

BREEDING
Breeding season: spring and summer

HABITAT
Larva: foliage of host trees in broadleaf woodland. Adult: may wander further afield.

DISTRIBUTION
Much of Europe and temperate Asia, east as far as Japan; also North Africa

STATUS
Uncommon

a small part of the case remains attached by hooks to the moths' head. This bears two sharply pointed spikes, which are used to cut a hole in the cocoon. After the moth has climbed through the hole, it pushes this cutting tool off its head with its legs. Cutting is made easy by the secretion of a weak solution of caustic soda from the mouth, which softens the tough wall of silk.

Red for danger

The caterpillar of the puss moth is remarkable for having two kinds of protective coloration. The pattern of brown and green disrupts the outline of the caterpillar, making it difficult to find. This is a passive form of protection. If disturbed, however, the caterpillar rears up, presenting a

terrifying picture; the head is drawn in and the next few segments are hunched, displaying a bright scarlet ring with two black spots above it. This display looks very much like a face and its sudden appearance may well deter predators. The scarlet ring is a warning. If further molested, the caterpillar produces formic acid from a special gland located in its throat.

Other moths of the family Notodontidae possess similar means of defense. Many of these moths have protective coloration, and one species secretes strong hydrochloric acid when disturbed. Even stranger in appearance than the puss moth caterpillar is the caterpillar of the lobster moth, *Stauropus fagi*. The lobster moth larva feeds primarily on beech trees, and is found from Britain east to Japan. The caterpillar is brown and has two pairs of very long, highly modified prolegs. The tip of the abdomen is raised vertically, and the rear portion is elongated and swollen in appearance. When it becomes annoyed, the lobster moth caterpillar rears up and waves its long prolegs in an attempt to frighten away the predator.

When under threat, the puss moth caterpillar retracts its head into the body and rears up to display a bright red "face" topped with two black eyespots.

PYROSOMA

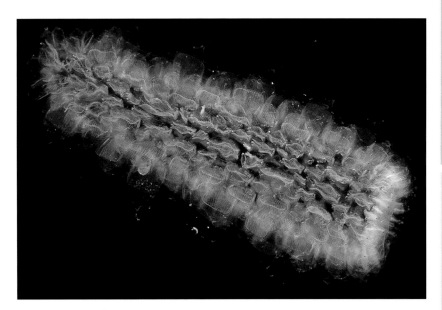

PYROSOMA	
PHYLUM	**Chordata**
SUBPHYLUM	**Urochordata**
CLASS	**Thaliacea**
ORDER	**Pyrosomida**
FAMILY	**Pyrosomidae**
GENUS	***Pyrosoma***
SPECIES	**Several, including *P. atlanticum***

LENGTH
From 4 in. to 9¾ ft. (10 cm to 3 m)

DISTINCTIVE FEATURES
A colony of up to many thousands of individuals, forming a jellylike cylinder

DIET
Microscopic organic particles

BREEDING
New colony formed by budding

LIFE SPAN
Not known

HABITAT
Open seas and oceans; normally down to 660 ft. (200 m)

DISTRIBUTION
All tropical and subtropical waters

STATUS
Widespread but not necessarily numerous

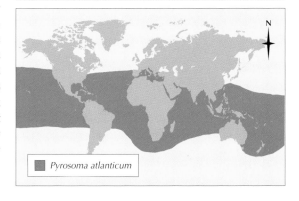

Pyrosoma atlanticum

Each individual member of a Pyrosoma *colony has two light organs that flash when the colony is disturbed or touched. The main predator of the colony is a small planktonic crustacean, genus* Phronima, *which eats the colony's living parts, leaving only the cylinder of jelly.*

FORMING FREE-SWIMMING colonies of sea squirts, the genus *Pyrosoma* is characterized by its cylinder shape that tapers slightly to a narrow, closed end. The cylinder may be about 4 inches (10 cm) to 9¾ feet (3 m) long and up to 10 inches (25 cm) in diameter. Its wall is made of a stiff jelly, almost like gristle, in which are embedded the individual sea squirts, numbering several thousands in the largest cylinders. Two other genera of group-living sea squirts, *Pyrostremma* and *Pyrosomella*, have similar features.

Sea squirts, or tunicates (subphylum Urochordata), form a group of marine animals that have been described as being among the ancestors of vertebrates. Of relatively simple structure, an individual sea squirt is enclosed in a tunic of a tough, jellylike substance. It has a mouth through which water is drawn in and a vent from which it is expelled. In between, the water passes through a cavity, the pharynx, the walls of which form a fine-meshed, mucus-covered network that acts as gills. Oxygen is taken from the water by the gills, and tiny particles of food are trapped by them and passed to the stomach.

Jet propulsion

The *Pyrosoma* cylinder moves using a form of jet propulsion. The beating of the gill cilia of each colony member draws in a stream of water, which is expelled through the opening at the cylinder's end, driving the colony along. The cylinder moves slowly, only slightly faster than the speed of the current in which it is drifting. The *Pyrosoma* species live in the open oceans of the Tropics, swimming well below the surface by day but coming up to the surface at night.

A *Pyrosoma* ovary contains one ovum heavily charged with food-yolk. When fertilized, the ovum divides repeatedly to form a multicellular embryo. From this a creeping stalk, or stolon, grows out and becomes constricted into up to 50 parts, each of which is the beginning of a new *Pyrosoma*. This group of "buds" develops a jellylike tunic and, with the embryo from which it sprang, leaves its parent and builds up a new colony by budding from its up to 50 new parents.

PYTHON

PYTHONS ARE THE EQUIVALENTS in Africa, Asia and Australia of the American boas. Like the boas, they have small spurs on their bodies that represent the vestiges of hind limbs. The largest and best-known pythons belong to the genus *Python*. The African python, *Python sebae*, the largest individual of which species was 26½ feet (8 m) in length, lives in most parts of sub-Saharan Africa. The other African pythons are the ball python (*P. regius*), often called the royal python, and the dwarf python (*P. anchietae*). There are no pythons in southwestern Asia, but several species are found from India to China and Indonesia. The Indian python, *P. molurus*, reaches about 20 feet (6 m), although the largest specimen ever recorded measured 29½ feet (9 m). The Indian python ranges from India through Southeast Asia to China and occurs on some Indonesian islands. The reticulated python, *P. reticulatus*, has a more easterly distribution, from Myanmar (Burma) to the Philippines and Timor. There has been much dispute about which is the biggest python, but the record is probably held by a reticulated python that measured 33 feet (10 m).

Many other snakes belonging to the family Boidae are also called pythons. There are 14 such species in Australia (some have restricted distributions and have never been given English names) and others in New Guinea and parts of Indonesia. The largest is the amethystine python, *Morelia amethistina*, which is found in New Guinea and northeast Queensland. The biggest specimen ever recorded was almost 28 feet (8.5 m) in length, but most specimens are less than half this size. The green tree python (*Chondropython viridis*), the carpet python (*Morelia spilota*) and the diamond python (a subspecies of the carpet python) also live in New Guinea and Queensland. The children's python (*Liasis childreni*) and the black-headed python (*Aspidites melanocephalus*) are found only in the northern parts of Australia. The woma (*A. ramsayi*), on the other hand, is confined to the arid interior of Australia. These snakes are usually active only at night, although individuals of almost all python species may bask in the sun by day after swallowing a large meal.

Varied habitats

Large pythons in Africa are often found near water, and the Indian python is almost semi-aquatic. Pythons also live in jungles and climb trees, except for the African python, which prefers open country, though not deserts. The reticulated python shows a preference for living near human settlements. At one time it was a regular inhabitant of Bangkok, hiding by day and coming out at night to feed on rats, cats, dogs and poultry. One individual was caught in the King's Palace. This habit of associating with buildings must account for its turning up in ships' cargoes. A reticulated python once reached London in good condition. This species is also a good traveler under its own steam. It swims out to sea and was one of the first reptiles to recolonize the island of Krakatoa, between Sumatra and Java, after it erupted in 1883.

A green tree python lies draped in a tree in Irian Jaya, Indonesia. This snake's leaf-green coloration with white spots along its back plus its extremely prehensile tail are superb adaptations for life in trees.

An African python devours a Thomson's gazelle, Gazella thomsoni, *in Amboseli National Park, Kenya.*

PYTHONS

CLASS	**Reptilia**
ORDER	**Squamata**
SUBORDER	**Serpentes**
FAMILY	**Boidae**
GENUS	**Several, including *Aspidites*, *Chondropython*, *Liasis*, *Morelia* and *Python***
SPECIES	**27, including green tree python, *Chondropython viridis*; amethystine python, *Morelia amethistina*; African ball python, *Python regius*; reticulated python, *P. reticulatus*; and African python, *P. sebae***

ALTERNATIVE NAMES
Green python (green tree python); Australian scrub python (amethystine python); royal python (African ball python); African rock python (African python)

LENGTH
Reticulated python: up to 33 ft. (10 m)

DISTINCTIVE FEATURES
Large, heavy body; 1 pair of small spurs on either side of vent; small pits on lips

DIET
Mammals; also birds, reptiles and amphibians

BREEDING
Age at first breeding: 5 years; number of eggs: sometimes more than 100; hatching period: 60–90 days

LIFE SPAN
More than 20 years in larger species

HABITAT
Very variable, from savanna to forest

DISTRIBUTION
Sub-Saharan Africa, India, Southeast Asia, Indonesia, New Guinea and Australia

STATUS
Larger African and Asian species vulnerable

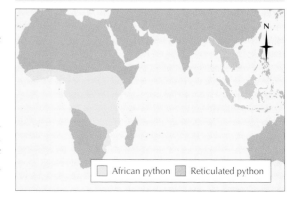

African python   Reticulated python

Any live prey accepted

Pythons kill their prey by constriction, wrapping themselves around the body of the prey so that it cannot breathe. The coils then hold the body steady while the python works it into its mouth. Prey is caught by ambush. The python lies in wait and then springs out, striking the animal with its head and seizing it with its jaws until it can wrap its body around it. The list of animals that pythons eat is too long to catalog. Pythons prefer mammals, followed by birds, but people have found young African pythons caught in fish traps. African pythons eat many small antelopes, such as duikers, gazelles, impalas and bushbucks, and a large python can swallow prey weighing up to 120 pounds (55 kg), although this is exceptional. Usually pythons take smaller animals, such as hyraxes, hares, rats, pigeons and ducks. Jackals and monkeys are sometimes eaten, and one African python 18 feet (5.5 m) long is known to have eaten a leopard, having sustained very little damage in the process of catching it. Pythons sometimes suffer from their meals. They have been found with porcupine quills and antelope horns sticking through the stomach wall. Such dangerous projections do not usually cause any serious damage, and are soon digested.

A meal of a large animal will last one of these snakes a long time, but a python sometimes kills several small animals, one after the other. An African python has been credited with capturing and eating three jackals, and a small python was seen to kill two sparrows in quick succession and then pin down a third with its tail. There are a number of authentic accounts of people being attacked and swallowed by larger pythons, both in Africa and Southeast Asia.

Like that of many snakes, the courtship of pythons can be quite elaborate. The male crawls after the female, trying to climb over her, and sometimes the two snakes rear up and sway to and fro. The little spurs which lie on either side of the cloaca may be used by the male to scratch the female and stimulate her to raise her body and bring the two cloacas together. Large pythons may lay large clutches of eggs. More than 100 have been recorded on a number of occasions. In many of the larger pythons, the female gathers the eggs into a pile and wraps herself around them, guarding them throughout the 2–3-month incubation period, leaving them only for occasional visits to water or more rarely to feed. Females of many species simply watch over the eggs in this way, but in others (for example, the Indian python) the females incubate the eggs by keeping their body temperature a few degrees above that of the surroundings. They do this by slowly contracting the muscles which run along the body. This is a primitive kind of shivering. Juvenile reticulated pythons are 2–2½ feet (60–75 cm) long when they hatch. They grow at a rate of about 2 feet per year and become sexually mature after about 5 years.

Even the great snakes are not free from predators. Young pythons are preyed on by many animals, but as they grow larger, fewer animals can overcome them. Crocodiles, hyenas and tigers have been found with the remains of pythons in their stomachs, and one naturalist writes of finding a 17-foot (5.2-m) Indian python killed by a pair of otters, which had apparently attacked from either side, avoiding harm by their agility. When Africa's ball python is molested, it rolls itself into a tight, almost uniformly round ball, with its head tucked well inside.

Beating elephants

Pythons seem to be responsible for one of the many dragon legends, the word *dragon* being derived from the Greek word for snake. In his *Historie of Serpentes* of 1608, Edward Topsell described how dragons capture elephants. He writes how they "hide themselves in trees covering their head and letting the other part hang down like a rope. In those trees they watch until the Elephant comes to eat and croppe off the branches, then suddainly, before he be aware, they leape into his face and digge out his eyes, and with their tayles or hinder partes, beate and vexe the Elephant, until they have made him breathlesse, for they strangle him with theyr foreparts, as they beat him with the hinder." Apart from the impracticability of an elephant being attacked, this is a reasonable account of a python killing its prey.

This front-on view of a reticulated python, the largest of all pythons, shows the pits between the scales on the lips that enable the snake to detect heat from its warm-blooded prey.

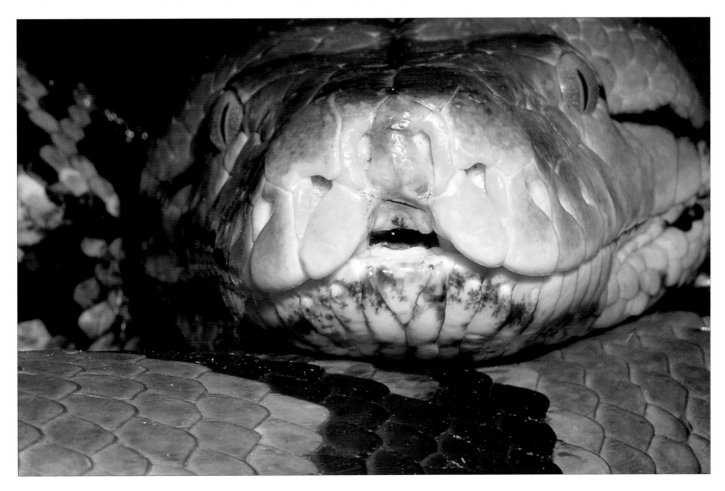

QUAIL

Like other members of the pheasant family, quails take seeds, grain, insects, small snails and other small invertebrates.

Slow off the ground

Quails are similar to partridges (discussed elsewhere in this encyclopedia) in habits, although they are even more reluctant to fly. Like partridges, they keep to arable land or pasture and avoid tall vegetation. When flushed, they fly low, 3–4 feet (0.9–1.2 m) from the ground for short distances. They gather in small family parties or bevies and even on migration form parties of no more than 40 or so. The male common quail's call sounds like "wet-my-lips," while the female's call is a double note, *brub-brub*.

Migratory common quails were formerly found in large numbers and at one time in the Middle East hunters knocked these low-flying birds down from the air using sticks. In medieval times large numbers reached northern Europe during the summer. In England in 1466, 100 dozen quails were served at a banquet given for the Archbishop of York, and quails were still abundant in 18th-century England. New World quails have similar habits, but none is migratory. The northern bobwhite, *Colinus virginianus*, is a popular game bird. Its common name is based on the male's call. The marbled wood quail, *Odontophorus gujanensis*, native to parts of South America, is remarkable for its duets: male and female stand facing each other about 12 inches (30 cm) apart, calling *corcorovado*; the male calls *corcoro* to which the female adds *vado*.

Perfunctory courtship

In the breeding season the male common quail takes up a territory from which he drives out other males. A female arrives, chooses a nesting site and then calls to the male, after which the two perform duets together. The male goes over to the female and circles her, dragging his wings, stretching out his neck and puffing out his throat. After mating, the female makes a shallow scrape in the ground, lines it with a few bits of grass and lays 8 to 13 eggs, which are yellowish white with amber to chocolate markings. Having laid her eggs, the hen incubates them for about 3 weeks. The young leave the nest within a few hours of hatching and can fly well at about 19 days. The breeding details for the northern bobwhite and related New World quails are similar, with the males taking a greater share in tending the young. Northern bobwhites lay a larger clutch on average, with 12 to 18 eggs.

Gambel's quail, Callipepla gambelii, lives in arid areas of the southwestern United States and parts of Mexico. It is also called the topknot quail, after the distinctive plume that extends from its head.

THERE ARE MANY SPECIES of quails, including 31 species of New World quails, 16 species of button quails in Asia, Africa and Australasia, and several other species of Eurasian and African quails in the genera *Coturnix* and *Perolicula*. Although New and Old World quails resemble each other superficially and are similar in habits, they are very different in anatomy.

The name quail came into common use between 1550 and 1620 and was undoubtedly taken to North America at that time by European settlers. The name subsequently declined in common usage, but revived in the mid-18th century.

The migratory common quail, *Coturnix coturnix*, is abundant across Europe, Asia and North Africa. It is plump, up to 7 inches (18 cm) long, with a very small tail and a weak bill and legs. The plumage is buff and brown mottled, barred and streaked with black, dark brown and white. Related quails of similar build, living in Africa, Southeast Asia, Australia and New Zealand, are more brightly colored.

New World quails are usually slightly larger and more brightly colored than Old World quails. The tail may be short or fairly long, and the bill is stronger, with serrated edges. As well as having crests of various shapes and sizes, the crested quails are the most colorful quail species.

COMMON QUAIL

CLASS	**Aves**
ORDER	**Galliformes**
FAMILY	**Phasianidae**
GENUS AND SPECIES	***Coturnix coturnix***

WEIGHT
2⁹⁄₁₀–4¾ oz. (75–135 g)

LENGTH
Head to tail: 6⅓–7 in. (16–18 cm)

DISTINCTIVE FEATURES
Small, plump bird; complex pattern of dark and pale stripes on head; brown body with buff streaks; very short legs

DIET
Mainly seeds, grain and insects

BREEDING
Age at first breeding: 1 year; breeding season: eggs laid April–August; number of eggs: 8 to 13; incubation period: 17–20 days; fledging period: 19 days; breeding interval: 1 or 2 broods per year

LIFE SPAN
Up to 8 years

HABITAT
Steppes and cultivated land

DISTRIBUTION
Breeding range: much of Europe, apart from far north; Central Asia south to India; northwestern, eastern and southern Africa; central Madagascar

STATUS
Common

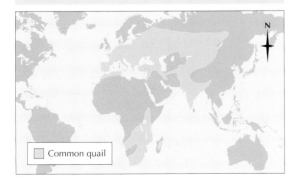

Common quail

Like most pheasants and other quail species, the common quail nests on the ground. This can leave the young highly vulnerable to attack from predators.

Like partridge coveys, quails roost in circles with their heads turned outward. Northern bobwhites use the circle as a defensive formation or burst dramatically from cover to confuse a predator. Even so, because they feed, sleep and nest on the ground, all quails are vulnerable to ground predators such as foxes and snakes, and they are also attacked by hawks and owls.

Arriving together

A northern bobwhite hen lays an egg a day until four or five are laid. She then may miss 2 or 3 days before laying the next. Consequently, a clutch of 18 eggs may take over 3 weeks to lay. However, at the end of the 23–24 days' normal incubation, all the eggs hatch within 2 hours. The same pattern applies to the gray partridge, ring-necked pheasant and willow ptarmigan. To a certain extent this is because the hen does not start incubating until the full clutch is laid, as watching out for straying chicks as well as incubating eggs would be a difficult task for her. It is clearly advantageous to the hen if all her eggs hatch at about the same time. Two days or so before the eggs hatch they are pipped; that is, the chick pecks a tiny hole and starts breathing with its lungs. A day later the chick starts to click: it makes regular sounds at a rate of 80 to 150 per minute. Synchronized hatching takes place only when the eggs are close together or almost touching. It seems that the most advanced eggs cause the less advanced to speed up their development and that the less advanced hold back the more advanced, making them mark time.

Crested quails have a slightly more elaborate courtship in which the male, which is more colorful than the female, ceremoniously presents his mate with food. Between the offerings he bobs up and down, whistles low and dances around her delicately before mating occurs.

QUELEA

Red-billed queleas roost, nest and feed together. They are superabundant because their food source, grass grain, is now widely cultivated by humans.

THE QUELEA IS A MAJOR agricultural pest, doing damage worth millions of dollars to crops in Africa every year. Its generic name *Quelea* has become accepted in the English language, but it is also known in English as the dioch. There are three species in the genus *Quelea*. The main pest is the red-billed quelea, whose name draws attention to the feature that best distinguishes it from the other two species. It is a small bird, rather like a sparrow, but with a stout, conical, red bill. In breeding dress, the male has a buff forehead, crown, nape and breast and a conspicuous mask on its cheeks. The mask is usually black, but in a number of males it is buff and barely noticeable. This is an example of dimorphism (see "Guillemot"), with the black-faced form making up 80–90 percent of the population. After the breeding season the male molts his brighter colored feathers and becomes brown and sparrowlike for several months. The female is always dull, resembling the male in his nonbreeding dress. Only her bill changes color: from red to bright yellow at the height of the breeding season.

The red-billed quelea is found throughout the dry savannas and grasslands of Africa south of the Sahara, a total range of over 2 million square miles (5,180,000 sq km).

Vast flocks

At times, queleas can be seen in quite incredible numbers on the rich floodplains throughout the semidesert regions of Africa. At other times, one can travel hundreds of miles across the same plains without seeing one. A single quelea is an unusual sight because they usually move about in tight flocks of hundreds to tens of thousands of birds. During the early morning, the birds feed intensively in dense clusters that form living carpets often 100 square yards (84 sq m) or more in area. They quickly fill their stomachs and crops, the crop bulging out like a balloon so the contents can be readily identified through the stretched skin of the neck. During the hot hours of the day, the flocks gather in some shady place, usually near water. There they sit chattering and preening and from time to time fly down in groups to drink on the edge of a pool. In the cool of the late afternoon, there is another feeding session. At dusk the flocks fly to the communal roost, which may be a patch of thorn trees or a reed bed, to spend the night. The roost can be colossal, with many millions of queleas packed into a few acres.

QUELEAS

CLASS	**Aves**
ORDER	**Passeriformes**
FAMILY	**Ploceidae**

GENUS AND SPECIES **Red-billed quelea, *Quelea quelea* (detailed below); red-headed quelea, *Q. erythrops*; cardinal quelea, *Q. cardinalis***

WEIGHT
About ⅗ oz. (18 g)

LENGTH
Head to tail: 4½–5 in. (11–13 cm)

DISTINCTIVE FEATURES
Sparrow-sized; conical bill. Male (breeding): rosy-red bill; black chin and cheeks which form a mask; yellow-buff underparts; mid-brown upperparts. Female and nonbreeding male: no face mask; duller bill.

DIET
Mainly grain and seeds

BREEDING
Age at first breeding: 1 year; breeding season: eggs laid at start of wet season; number of eggs: 3; incubation period: 10–12 days; breeding interval: often 2 years

LIFE SPAN
Not known

HABITAT
Bushy grassland, savanna and fields

DISTRIBUTION
Much of Sub-Saharan Africa apart from areas of true desert, rain forest and mountains

STATUS
Common to superabundant

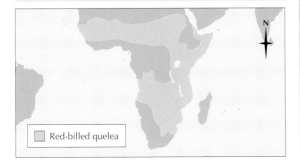

Red-billed quelea

In their daily activities the flocks may move 30–40 miles (48–64 km) from the roost, but sometimes queleas fly much greater distances. Birds ringed in South Africa have been recovered a thousand miles away in Malawi. With their food depending on seasonal rains, and the wet season coming at different times in different regions, it is not surprising that queleas sometimes fly long distances in search of food.

Stripping fields of grain

The natural food of the quelea is the tiny seeds of wild grasses. During the rainy season the seeds are stripped from the growing grasses, but in the longer dry season they are picked up off the ground. Queleas treat cultivated grasses, the cereal crops, just as they do the wild ones. Enormous numbers of queleas assemble in crop-growing areas when the rice, wheat, or guinea corn is ripening. By constantly feeding on these crops before the harvest, they can take a great deal of the grain and sometimes completely strip the fields. Peasant farmers crack whips, bang drums and shout for all they are worth to deter the birds. All they can do is to make sure the damage is shared equally by all, as the queleas merely fly to the next field. They still return to the roost every night with their crops bulging with grain.

Large rice- and wheat-growing schemes are now being developed all over Africa. These crops mature later than wild grasses, so extending the summer feeding season and decreasing the need for queleas to travel to find food. As much as 10–20 percent losses of crops are not now uncommon, and in some places one-quarter of the crop has disappeared down the birds' gullets before it can be harvested.

The red-billed quelea is also known as the avian locust. Flocks can total hundreds of thousands, and they can have a similarly destructive effect on crops.

Red-billed queleas can cover large distances. The birds travel en masse, in search of their locally abundant food supply.

Closely packed families

Queleas probably breed only once a year. At the beginning of the rains, the birds come into breeding condition. At the height of the rains they assemble in suitable breeding places. Often a line of acacia trees on the edge of a large swamp is selected or they may choose to build in the swamp itself. The nests are packed very close together and trees only 30 feet (9 m) tall may carry several hundred nests, while a single large colony, sometimes covering several square miles, can hold millions of nests.

The males begin the building and attract a female by displaying on the half-finished nest. After mating, the nest is quickly completed by the two partners. It is woven from strips of green grass and is a roughly oval structure with an entrance hole on one side. The eggs are often laid even before the nest is finished. The nestlings are fed on caterpillars, grasshoppers and other insects for a few days and grow very rapidly; then they are given grass seeds. After about 2 weeks, the young leave the nests, finally flying away when about 4 weeks old.

Few predators

One might expect the swarms of queleas to attract large numbers of predators, but this is not so. Falcons and goshawks can often be seen harrying the flocks, but when threatened, the queleas bunch together in close formation, and the dense mass of birds quickly twists and turns. All the hawks can do, it seems, is to catch the odd sick or injured bird that gets left behind. Humans take queleas for food, and some birds drown when flocks drink at shrinking water holes.

Is there no end to them?

For many years, control units have been killing queleas all over Africa. The present toll runs into millions each year, yet there are apparently still as many queleas as ever. The proportion of the total population killed is quite small and can be replaced even by their relatively slow rate of reproduction. If nests or young queleas are destroyed, the adults start to breed again almost immediately. Many of the breeding places are completely inaccessible to the control units and many others still remain to be discovered.

QUETZAL

THE QUETZAL MUST RANK as one of the world's most beautiful birds. It measures 15–16 inches (38–40 cm) from bill to tip of tail, but, like the peacock, the male's tail coverts are extremely long, trailing behind the quetzal as shimmering plumes up to 2 feet (60 cm) beyond the tail tip. The female shows many of the same features as the male, but she is duller, lacks a crest and her tail coverts are much shorter.

Resplendent quetzals are found in the forests of Central America, from southern Mexico to Panama, but are now becoming increasingly rare. Other species of quetzals are found in the northern Andes and west of the Amazon Basin.

Mind my tail

The quetzals live among the rich profusion of plants in the canopy of tropical cloud forest and rain forest, feeding on the fruits and berries. Flying from its perch, a quetzal takes a berry in its bill, tears it off and returns to its perch, so feeding in an action reminiscent of flycatchers. The male quetzal has an unusual habit of launching itself backward from its perch. If it flew straight from its perch in the normal manner, the long feathers in its train would be torn to shreds in no time. The female does not need to take this precaution.

Quetzals usually stay in the dense forests but sometimes come out to feed in clearings. In the forest the quetzals give away their position by their habit of rapidly fanning their tails to show flashes of white. They also call loudly in flight, and during the breeding season the males have a variety of deep, powerful calls and songs that match their beautiful plumage. The males also have a display flight, circling above the forest canopy and calling loudly.

Inaccessible nests

Quetzals nest in holes in tree trunks, either enlarging existing woodpecker holes or excavating their own. Their bills are short and blunt, so quetzals can make their holes in only the most decayed trees. Both male and female take part in nest building, biting and tearing strips of soft wood until they have made a hole with an entrance about 4 inches (10 cm) across and up to about 1 foot (30 cm) deep.

The light blue eggs are laid on the unlined floor of the hole and incubated by both parents in turn. According to Guatemalan folklore, the quetzal's nest has two entrances so that the male may sit on the eggs without ruining his train. He is supposed to enter through one side, incubate

with his train hanging out and then depart through the other side. In Costa Rica the quetzal is said to sit on the eggs with his head facing inward and his train trailing out through the single entrance. In fact, the train is sufficiently hardy to dispense with such elaborate behavior. The male quetzal incubates facing the nest hole

Among the Aztecs, quetzal feathers were worn only by nobles. The plumes were plucked from live birds, which were freed to grow new ones.

with the tail pressed vertically against the back wall. The train curves over his back and projects several inches through the entrance. Few people have seen quetzal nests because of the inaccessibility of the forests and the difficulty of locating nest holes in trees covered with a rich growth of epiphytic plants.

The chicks are naked and blind at first and for the first few days are brooded by their parents, who also keep the nest clean by removing the eggshells and droppings. The chicks are fed on insects and other small animals until they are nearly 2 weeks old. Then fruits, and later snails, small frogs and lizards, are brought by the parents.

Considering the quetzal's beauty, it is unsurprising that it is the national bird of Guatemala, where it has also given its name to the unit of currency. A pretty story has it that a captive quetzal soon dies of a broken heart. Such reverence for the quetzal has not saved it from the feather trade. After its discovery by Europeans, it soon became so rare that some people declared it was a mythical bird, but it was rediscovered in the 19th century, and only the inaccessibility of its home in the forests saved it from extinction.

Not only is the male resplendent quetzal the proud possessor of long tail coverts, he also has long wing coverts, which are also iridescent green and overhang the crimson lower breast and belly.

RESPLENDENT QUETZAL

CLASS	**Aves**
ORDER	**Trogoniformes**
FAMILY	**Trogonidae**
GENUS AND SPECIES	***Pharomachrus mocinno***

ALTERNATIVE NAME
Resplendent trogon

WEIGHT
About 7 oz. (200 g)

LENGTH
Head to tail: 15–16 in. (38–40 cm); including male's tail coverts: 15–40 in. (38–100 cm)

DISTINCTIVE FEATURES
Breeding male: very long, green tail coverts; small, tufted green crest. Female: duller; lacks crest; much shorter tail coverts.

DIET
Fruits; also insects, small frogs and lizards

BREEDING
Age at first breeding: 3 years (male); 1 year (female); breeding season: eggs laid March–June; number of eggs: 2; incubation period: 17–18 days; fledging period: about 21 days; breeding interval: 1 year

LIFE SPAN
Not known

HABITAT
Canopy of cloud forest and rain forest

DISTRIBUTION
Southeastern Mexico south to Panama

STATUS
Scarce or uncommon

Resplendent quetzal

QUOKKA

THE QUOKKA IS A WALLABY that is barely as large as a hare and resembles a large rat. It is also known as the short-tailed pademelon; its aboriginal name is *quak-a*. Head and body together measure up to 2 feet (60 cm), and the adults weigh up to 11 pounds (5 kg). The 14-inch (35-cm) tail is twice the length of the head, and the hind feet are only 4 inches (10 cm) long. The short, coarse fur is brownish gray with a reddish tinge around the head and chest.

Quokkas live in swamps and brush in south-western Australia. On nearby Rottnest Island they live in a variety of habitats. They were first seen on this island in 1658 by a Dutchman, Samuel Volckersen, who described them as "a wild cat resembling a civet cat but with browner hair." In 1696 another Dutchman, Vlaming, thinking quokkas were giant rats, named the island Rottnest (Rat's nest) Island after them. In 1830 a dead quokka was examined and found to be a type of wallaby.

Sheeplike quokkas

Quokkas live in dense grass with a moderate amount of scrub and scattered trees, making runways and tunnels through the vegetation. Although the hind legs of quokkas are short, they hop like kangaroos when going fast. When moving slowly, they do not use the tail as a prop like other wallabies do. They feed mainly at night, resting in the shade during the heat of the day and sleeping in small groups. They feed on grasses and herbaceous plants. During dry summers they search for succulent plants and damp places to obtain water.

Like rabbits, quokkas crop the vegetation and prevent regeneration of trees by eating the seedlings, but they resemble sheep in their digestion. They are semiruminants, possessing several kinds of bacteria in their stomachs to break down tough plant fibers.

Baby in reserve

Mating takes place throughout the year in some areas and from January to March in others. The gestation period is 26–28 days. The single newborn joey leaves the pouch at the age of 6 months but may still go back into it for shelter and rest until the youngster becomes fully independent at 10 months. Female quokkas mature sexually at 8½ months and males mature at 13 months.

Quokkas have delayed births. If the joey (young) that is already in the pouch and attached to a nipple is removed or dies, another joey will be born without further mating by the mother because female quokkas can become pregnant soon after their firstborn. The fertilized egg divides repeatedly to form a tiny hollow sphere called a blastocyst. So long as one joey is suckling, nothing more happens. The blastocyst lies quiescent in the uterus. If the first joey is lost, the blastocyst begins to develop and another joey is born after about 27 days. This type of delayed birth occurs in several species of marsupials.

Mystery of crash deaths

Although aboriginal people killed quokkas in small family groups, in the early days of the European settlement of Australia the animals were still plentiful. Quokka numbers were not seriously reduced until 1921 and 1933, when there were very high death rates. The causes are not known. Introduced rabbits and foxes had not reached that region by 1921, although both had arrived by 1933.

After the rise in quokka deaths in 1921, the populations readily built up again. This did not happen after the 1933 crash because rabbits took over much of the grasslands and foxes were present in sufficient strength to prey on the quokkas. Today, quokkas are most numerous in

Shade is essential to keep quokkas cool and stop them from licking themselves too much. Excessive licking means they lose vital minerals in their saliva, which affects their ability to digest plant fibers.

Too many quokkas in one area means seedlings are eaten before they can grow into bushes or trees and provide vital shade. This may explain the sudden falls in quokka populations witnessed in 1921 and 1933.

QUOKKA

CLASS	**Mammalia**
ORDER	**Diprotodontia**
FAMILY	**Macropodidae**
GENUS AND SPECIES	***Setonix brachyurus***

ALTERNATIVE NAME
Short-tailed pademelon

WEIGHT
4½–11 lb. (2–5 kg)

LENGTH
Head and body: 19–24 in. (48–60 cm); tail: 10–14 in. (25–35 cm)

DISTINCTIVE FEATURES
Plump, wallaby-like body; short, rounded ears; coarse, gray-brown or rufous fur; sparsely furred tail; short hind legs

DIET
Many different species of plants

BREEDING
Age at first breeding: 8½ months (female), 13 months (male); breeding season: usually January–March (Rottnest Island), all year (mainland); number of young: 1; gestation period: 26–28 days; breeding interval: about 1 year

LIFE SPAN
Not known

HABITAT
Areas of scrub and light woodland on offshore islands; also swampy woodland on mainland

DISTRIBUTION
Mainly on Rottnest Island and Bald Island, off coast of southwestern Australia; small populations on adjacent mainland

STATUS
Vulnerable; can reach high densities locally, such as 10,000 on Rottnest Island

the reserve set aside for them on Rottnest Island, where they are protected. They numbered approximately 10,000 a few years ago. Quokkas are now rare on the mainland.

Starving by not drinking

The quokka leads a biologically precarious life, partly because of its need for water and partly because it is a semiruminant. It keeps cool in hot weather by licking its forelegs and chest. The evaporation of this water lowers its body temperature. However, unless it drinks plenty of water, too many mineral ions are lost in its saliva. These ions are needed to maintain the stomach acidity necessary for the bacteria that break down the plant fibers. A quokka needs only about 1 ounce (28 g) of protein a day to keep its diet balanced, which it obtains from the fermentation of plant fibers in its stomach and by digesting the bacteria. It can also retain nitrogenous waste instead of excreting it. This passes into the stomach, where the bacteria convert it to fresh protein. The quokka can therefore make do with meager supplies provided it does not become too thirsty.

RABBIT

THE RABBIT USED TO be classed as a rodent (order Rodentia) but is now placed in the Lagomorpha, along with hares and pikas. It has long ears and large prominent eyes placed well to the sides of the head. The latter feature is common in herbivores. It gives panoramic vision as an aid to detecting predators.

The rabbit's strong hind legs are longer than the forelegs and provide the main force in running. The soles of the feet have a thick coating of hair, which gives a firm grip on either hard rock or slippery snow. The tail is white below, very short and turned up at the end. The coat is a mixture of brown and gray on the upperparts with creamy or white underparts and is made up of three kinds of fur. There is a dense, soft underfur, through which project the longer, stronger hairs that give the coat its color. Among these are still longer but more sparsely scattered hairs. The bucks are slightly larger than the females and may be up to 20½ inches (52 cm) in total length and weigh up to 5 pounds (2.3 kg).

Rabbits have been extensively bred for food, as pets and for laboratory experiments. Black rabbits and, more rarely, animals of other colors turn up occasionally in the wild, and these have been used to produce domesticated breeds, some of which bear little resemblance to the original wild stock. Domestic rabbits exist in a wide range of colors, and the fur of some breeds is very long and silky. Rabbits bred for their meat may reach a weight of 16 pounds (7.25 kg).

Originally a native of southwestern Europe, the rabbit has spread, largely with human help. It was introduced into many parts of the world, including the Ukraine, the Azores, Madeira, Australia, New Zealand, South America and several parts of the United States. In some places where it has been introduced, the rabbit has no natural predators and consequently has caused widespread devastation to grasslands and robbed the native animals of their food, particularly in Australia.

Underground communities

Rabbits prefer sandy soils and live mainly in grasslands or open woodlands, where they dig extensive burrows. They are gregarious and have their burrows close together, so a warren may cover a wide area. Although famous as diggers, rabbits are not especially built for burrowing. Yet where the soil is light the combined efforts of many generations have resulted in extensive and complicated systems of tunnels with bolt runs as

A group of young rabbits mill outside the entrance to one of their burrows.

RABBIT

CLASS	**Mammalia**
ORDER	**Lagomorpha**
FAMILY	**Leporidae**
GENUS AND SPECIES	***Oryctolagus cuniculus***

ALTERNATIVE NAMES
Old World rabbit; domestic rabbit

WEIGHT
2¾–5 lb. (1.3–2.3 kg); domestic breeds up to 16 lb. (7.25 kg)

LENGTH
Head and body: 13¾–17¾ in. (35–45 cm); tail: 1⅖–2¾ in. (4–7 cm)

DISTINCTIVE FEATURES
Light brown-gray or dark gray upperparts; creamy or white underparts; large ears

DIET
Grasses and herbs; also bark and vegetables

BREEDING
Age at first breeding: 3–4 months; breeding season: autumn–spring (original native range); number of young: 1 to 9, average 5; gestation period: 28–33 days; breeding interval: up to 5 litters per year in native range, less in drought or cold years

LIFE SPAN
Up to 3 years

HABITAT
Grasslands, fields, woodland meadows, parklands and scrub; mainly on sandy soils

DISTRIBUTION
Native to Iberian Peninsula and southwestern France; feral rabbits present in Eurasia, North Africa, North and South America, Australia and New Zealand

STATUS
Very common or abundant

Although mainly nocturnal, rabbits do feed during the day if undisturbed. Whatever the time of day, they rarely venture more than a few hundred feet from their burrows, although they may stray further on dark nights than on moonlit evenings.

emergency exits and stops, areas for nursery use. Although rabbits prefer the light sand of dunes or a sandy heath overgrown with furze and heather, they also drive tunnels into firm loam or dry clay. They have also been known to burrow deeply into coal. The forepaws are principally used to loosen the earth, which is then kicked back with the hind feet. If they meet stones that cannot be loosened using the paws, rabbits have been known to remove them with their teeth. A typical tunnel is about 6 inches (15 cm) in diameter, with passing places 12 inches (30 cm) wide at points along its length. The living quarters are always blind chambers leading from the main passages. Adult rabbits use no bedding materials but rest on the bare soil. Rabbits are mainly nocturnal, coming out of their burrows in the evening and returning in the early morning.

Although they occasionally eat snails or earthworms, rabbits are almost exclusively vegetarian, their chief food being grasses and the tender shoots of furze. In winter, when herbs are not available, rabbits eat tree bark. They can devastate crops of vegetables, and before the disease myxomatosis drastically cut numbers, rabbits inflicted heavy losses on farmers. A rabbit voids two types of droppings. One kind is eaten again, a process known as refection, or coprophagy. The other is discarded at a special latrine outside the burrow.

Prolific breeders

Rabbits are not promiscuous breeders but are polygamous, one buck mating with several does and each doe keeping to her own territory within the warren. Mating is preceded by a courtship in which the buck chases the doe. The gestation period is 28–33 days, and in the rabbit's native range, litters of babies succeed one another rapidly at intervals of about 1 month from autumn to spring.

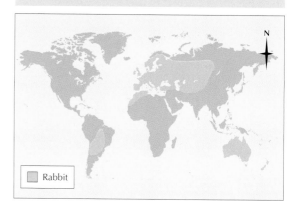

Rabbit

The pregnant doe makes a bed for her young from hay and straw lined with fur stripped from her underparts. The nest is usually in the blind end of a stop, or stab, a short burrow about 2 feet (60 cm) long, just under the surface and well away from the main burrow. Raising the young in a separate tunnel is a necessary precaution against the young being killed by the dominant buck, especially if he is not the father. The doe visits the nest to suckle the young for about 3 minutes once in every 24 hours. The entrance to a stop is closed with earth after each visit, and some does spread dried grass over the entrance to hide it.

Litters vary in size from two or three young up to nine, the larger litters being born in the warmer months. The newborn rabbits are blind, deaf and almost naked. Their ears are closed and they have no power of movement until about the tenth day, the eyes opening a day later. In a few more days the young rabbits can run and make short excursions from the nest. They start taking solid food at 16 days and are weaned after 30 days, when they are capable of an independent existence. Until then, the mother defends them from predators, using her powerful hind feet as weapons. The weight of a young rabbit at birth is 1–1¼ ounces (28–35 g), which increases to 9½ ounces (270 g) in 3–6 weeks. Sexual maturity is reached in 3–4 months. The number of young born does not correspond to the number of eggs fertilized. In recent years it has been established that a percentage of the embryos are resorbed, which means there is a degeneration of the embryonic tissues, the substance of which is taken back into the mother's body through the wall of the uterus. Whole litters sometimes are resorbed.

Thumping on the ground

The rabbit's chief predators, in addition to humans, are the members of the weasel family (Mustelidae), rats, owls, buzzards, ravens, crows, black-backed gulls and a variety of hawks. Badgers dig out the young, and foxes also take a large toll, as do cats and dogs. When a rabbit is disturbed, it usually makes an alarm signal, generally by thumping the ground with both hind feet together. All rabbits within earshot respond to the alarm by dashing toward their burrows. A rabbit, terror-stricken by the imminent attack of a stoat, utters a loud scream, and females have also been heard to utter low notes when nursing her young. Otherwise rabbits are normally silent, except for occasional low growls and grunts.

Results of myxomatosis

In 1954 and 1955 an introduced disease called myxomatosis made the wild rabbit a rare animal in western Europe. The disease diminished in severity and the rabbit population recovered. However, as soon as numbers begin to build up in any locality, the disease seems to take its toll once again. More recently a new disease, called calcivirus, has also spread.

In Great Britain, the changes following the first wave of myxomatosis highlighted the effects that rabbits' feeding had on the countryside. In a year or two after the drastic reduction in rabbit numbers of green lanes became choked with long grass, seedling trees and brambles, while the margins of cultivated fields were not being eaten bare. On downlands not being cropped by sheep the grass grew taller. At the same time many woodlands recovered and many wild flowers became more plentiful, wild orchids in particular becoming more abundant.

A report published in 1960 described the next stage. Unrestricted growth of the grass led to turf becoming less dense. As a result there was a reduction in the numbers of wild flowers that had so suddenly burst into prominence as the numbers of rabbits fell. There was an increase in brambles, gorse and heather. The report stated that the vegetation and scenery had reverted to its pre-1840 condition. It seems that until then relatively few rabbits lived outside the carefully guarded warrens where they had been preserved in a state of semidomestication since their introduction into Britain in the 12th century.

A rabbit has glands under its chin that produce a secretion used for marking territory. Rabbits also rub their chins on each other, especially on a mate or on the young, to provide a ready means of recognition.

RABBITFISH

When they are not browsing on coral reefs, rabbitfish rest in midwater at depths of 6½–20 feet (2–6 m).

THE 27 SPECIES OF rabbitfish in the family Siganidae derive their name from their rounded noses and their rabbitlike teeth. Rabbitfish have unusual reticulate markings on their bodies. These markings may be spots or mottlings but more often they are made up of irregularly curving lines that, when closely set, look more like a maze pattern. This could be described as scribbling but should be properly called vermiculate.

Rabbitfish can grow to 21 inches (53 cm) long in the case of the streaked spinefoot, *Siganus javus*. Usually, however, they are 1 foot (30 cm) or less in length. They have almost oval bodies, somewhat flattened from side to side, with a short tail. The dorsal fin runs the length of the back, and its front three-quarters is supported by 13 stout spines, the first of which points forward. The anal fin is long, and its front two-thirds is supported by seven stout spines. Each pelvic fin has five spines: the outer two are stout, the three between are soft. Each stout spine is grooved and has a poison gland at its base. The spines can inflict a painful but not dangerous wound.

Rabbitfish live inshore, ranging from the Red Sea to the coast of South Africa, east across the Indian Ocean to the Pacific, southward to the Great Barrier Reef off Australia and northward to Japan. The vermiculated spinefoot, *S. vermiculatus*, is the only truly estuarine species. The orange-spotted spinefoot, *S. guttatus*, prefers a low-salinity environment, and the little spinefoot, *S. spinus*, is also found in rivers.

Rabbitfish are noted for their ability to change color rapidly, a common feature of coral reef fish. They are mainly herbivorous, browsing the small seaweed on corals and rock surfaces, biting off a certain amount of animal food, such as coral polyps, at the same time. Rabbitfish are sometimes eaten by humans, although certain species are known to cause food poisoning.

Artificial fertilization
All rabbitfish are pelagic spawners. The eggs are colorless and about 2 millimeters in diameter. They hatch in 27 hours, the larvae being about 2.5 millimeters long and without visible eyes, mouth or pectoral fins; the larvae can only

wriggle their bodies and feed on the contents of the yolk sacs. Their pectoral fins start to grow 20 hours after hatching, and in 48 hours the eyes are black, the mouth is open and the larval fish is ready to feed itself.

From one sea to another

The Suez Canal, which connects the eastern Mediterranean with the northern Red Sea, has posed questions other than those concerned with international navigation. One of these problems, which has attracted the attention of scientists for many years, is whether or not marine animals would pass from the Mediterranean Sea to the Red Sea and vice versa.

The fauna (animal life) of the eastern Mediterranean is extremely limited compared with the western half of that sea, and it is also poorer than that of the Red Sea. At the southern end of the 100 miles (160 km) of the Suez Canal the water is 1 foot (30 cm) higher than at the Mediterranean end. For 9 months of the year the flow is from the Red Sea northward, which would suggest a migration of Red Sea species into the eastern Mediterranean. However, in 1929 Walter Steinetz found that only 15 species, including shrimps, oysters and fish, had gone north and that none had migrated southward.

One barrier was the Great Bitter Lake, through which the canal passes. This lake had a layer of salt on its bed estimated at 1,000 million tons (900 million tonnes), although the salt has been gradually dissolving. At the same time, the Aswan High Dam has made a difference to the flow of fresh water down the Nile to the eastern Mediterranean. One result of the dam's construction is that the water in the seas at either end of the canal are becoming more alike, as is the water passing through the Suez Canal. Scientists have discovered that several species of rabbitfish have gotten through from the Red Sea and that some have penetrated south from the Mediterranean Sea. Two species that reached the Mediterranean were the marbled spinefoot, *S. rivulatus*, and the dusky spinefoot, *S. luridus*.

RABBITFISH

CLASS	**Osteichthyes**
ORDER	**Perciformes**
FAMILY	**Siganidae**
GENUS	***Siganus***
SPECIES	**27, including streaked spinefoot, *Siganus javus* (detailed below), and vermiculated spinefoot, *S. vermiculatus***

LENGTH
Up to 21 in. (53 cm)

DISTINCTIVE FEATURES
Oval, compressed body; rounded nose; stout spines on dorsal and anal fins; delicate vermiculate markings

DIET
Algae

BREEDING
Poorly known

LIFE SPAN
Not known

HABITAT
Shallow coastal waters, brackish lagoons and rocky coral reefs

DISTRIBUTION
Indian Ocean and South Pacific

STATUS
Common

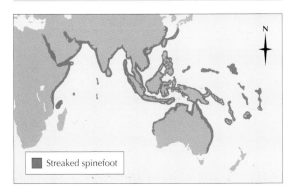

Streaked spinefoot

A masked rabbitfish, Siganus puellus. *The numerous spines on the dorsal and anal fins of rabbitfish are a protective feature: poison glands lie at the base of each stout spine.*

RACCOON

RACCOONS ARE ONE OF the best known North American animals, if only because of their appearance in folklore and stories. Their adaptability has allowed them to withstand drastic changes in their native environment. Raccoons have a head and body length of 16–24 inches (41–60 cm), with a tail of 8–16 inches (20–41 cm), and they weigh up to 26½ pounds (12 kg). Their fur is gray to black with black rings on the tail and a distinctive black "burglar mask" over their eyes. Their feet have long toes, while the front paws are almost handlike and exceptionally dexterous.

Raccoons are relatives of pandas, kinkajous and coatis. There are six species, the best-known of which is the common or North American raccoon, *Procyon lotor*, which ranges from Canada to Central America. The crab-eating raccoon, *P. cancrivorus*, lives in much of South America, from Venezuela south to Argentina. The other raccoon species are the Tres Marías raccoon (*P. insularis)*, the Cozumel Island raccoon (*P. pygmaeus*) and the Guadeloupe raccoon (*P. minor*), all of which are are found on islands in the Caribbean. The Bahama raccoon (*P. maynardi*) is critically endangered, while a seventh species, the Barbados raccoon (*P. gloveranni*) is now believed to be extinct.

Raccoons often respond to danger by raising the fur and arching the back, making themselves appear larger and more threatening to predators.

An adaptable species

Raccoons originally lived in woods and brush country, usually near water, but as the woods were cut down, they successfully adapted to life in open country. Raccoons are generally solitary animals, each one living in a home range of about 40 acres (16 ha), with a den in a hollow tree or a rock crevice. Occasionally they live in extended families and share dens, although it is rare for more than one adult male to belong to a den. Raccoons are mostly nocturnal creatures, although in some areas they may be diurnal (day-active), and are good climbers and swimmers. In the northern part of their range raccoons grow a thick coat and sleep through cold spells. The raccoons of the southern United States and southward are active throughout the year; they are also smaller than those in the north.

Where trees have been cut down raccoons move into fox burrows or barns, and they have been known to spread into towns, even to the middle of cities, where they live in attics and sheds and raid garbage bins for food. This last characteristic is one of the raccoons' less popular traits. Apart from the mess they cause, raccoons sometimes carry entire bins away so that they can search them at their leisure. There are stories of ropes securing the bins being untied by

RACCOONS

CLASS	**Mammalia**
ORDER	**Carnivora**
FAMILY	**Procyonidae**

GENUS AND SPECIES   **North American raccoon,** *Procyon lotor;* **crab-eating raccoon,** *P. cancrivorus;* **Tres Marías raccoon,** *P. insularis;* **Bahama raccoon,** *P. maynardi;* **Cozumel Island raccoon,** *P. pygmaeus;* **Guadeloupe raccoon,** *P. minor*

LENGTH
Head and body: 16–24 in. (41–60 cm); shoulder height: 9–12 in. (23–30 cm); tail: 8–16 in. (20–41 cm)

DISTINCTIVE FEATURES
P. lotor. Gray fur, often with red tinge; 4 or 5 black rings on tail; black eye stripes; sparsely furred paws; long toes.

DIET
Fruits, birds, eggs, fish, frogs, small mammals and invertebrates; also carrion and scraps

BREEDING
Age at first breeding: 1–2 years; breeding season: December–August (*P. lotor*), July–September (*P. cancrivorus*); number of young: usually 4; gestation period: 60–75 days; breeding interval: 1 year

LIFE SPAN
Up to about 5 years

HABITAT
Scrub and forest; also suburban areas (*P. lotor*)

DISTRIBUTION
P. lotor: Canada south to Panama; introduced to northeastern Europe. *P. cancrivorus:* much of South America. Other species: Caribbean.

STATUS
P. lotor, P. cancrivorus: common. *P. minor, P. insularis, P. pygmaeus:* endangered. *P. maynardi:* critically endangered.

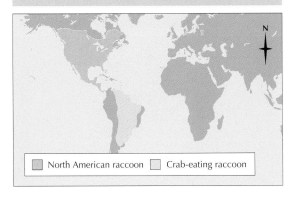

| North American raccoon | Crab-eating raccoon |

raccoons rather than being bitten through, an example of their dexterity. Experiments have shown that raccoons' sense of touch is very well developed and that they are almost as skillful with their hands as monkeys.

Raccoons are agile climbers and spend nearly as much of their time in trees as they do on the ground.

A wide-ranging diet

Raccoons eat a very wide variety of both plant and animal food and it is the ability to take so many kinds of food that is probably the secret of their success and explains their ability to survive changes in the countryside. Raccoons are primarily carnivores; earthworms, insects, frogs and other small creatures are included in their diet, and they also search in swamps and streams for crayfish and along the shore for shellfish. The eggs and chicks of birds, both ground- and tree-nesters, are eaten, and raccoons are sometimes pests on poultry farms and in waterfowl breeding grounds. They are also pests on agricultural land, due to the fact that they invade fields of corn, ripping off the ears and scattering them, half eaten. Fruits, berries and nuts are also taken.

Mother rears the young

Raccoons mate from December to August, each male mating with several females and then leaving them to raise the family. The young, usually about four in a litter, are born from April to June, after 60–75 days' gestation. Newborn raccoons weigh 2½ ounces (71 g) and are clad in a coat of fuzzy fur, already bearing the characteristic black raccoon mask. Their eyes open in 18 days, and at about 10 weeks they emerge from the nest for short trips with their mother. These trips become longer as the young learn to forage for themselves, but they stay with their mother until they are about 1 year old. Raccoons have lived up to 20 years in captivity, although in the wild they generally live for about 5 years.

Coonskin currency

Raccoons are a match for most predators and when hunted with dogs they may come off best, especially if they can lure their pursuer into water, where it may drown. Raccoons have been trapped and hunted in large numbers for some time in North and South America, first by Native Americans and subsequently by European settlers. The animals were hunted primarily for their hard-wearing fur, but also because of their attacks on crops. Even in the 17th century taxes and bans were imposed to prevent too many raccoon pelts from being exported. At one time the skins were used as currency, and when the frontiersmen of Tennessee set up the State of Franklin, the secretary to the governor received 500 coonskins a year, while each member of the assembly acquired three a day. Nowadays coonskin is not widely considered valuable.

Why so fastidious?

In the *Systema Naturae*, the Swedish naturalist Carl Linnaeus called the raccoon *Ursus* (later *Procyon*) *lotor*, or the "washing bear." The animal's common name originates from the Native-American word *aroughcan*, or *arakun*, meaning "one that scratches with its hands." It is similarly named in other languages: *ratons laveur* in French, Spanish *ositos lavadores* in Spanish and *Waschbaren* in German. These names testify to the habit raccoons have of apparently washing their food before eating it.

Some books state that raccoons always wash their food, whereas others say that the habit may be more common in captive animals. The first scientific study of food washing was made by a scientist at London Zoo, who showed that raccoons do not really wash their food but immerse it, manipulate it and then retrieve it. He suggested that the habit should therefore be called dousing.

The scientist gave a large variety of foods to a number of raccoons. Animal food was doused more often than plant food, yet earthworms, the only food that needed cleaning, were doused least of all. In another series of experiments it was shown that the shape, smell and size of food objects governed dousing to some extent, but that the most important criterion was the distance of the food from water. The nearer the water, the more likely the raccoon is to douse its food. Scientists now believe that raccoons wash food that they consider to be dirty before they consume it, and also that they remove the fleas from mammalian prey by immersing it in water.

Raccoons are born with their eyes closed and covered in fuzzy fur. They are weaned and start to hunt for themselves after 2 months, but stay with their mother for up to 1 year.

RACCOON DOG

THE RACCOON DOG IS a little smaller than the red fox, *Vulpes vulpes*, with a head to tail length of about 3 feet (90 cm), but its body is much stouter and its legs are shorter. It derives its name from the patch of black around and under each eye that resembles the "robber mask" of the raccoon. The fur is gray or creamy brown, with dark hair forming a collar over the shoulders, at the tip of the bushy tail and on the legs. The fur is long, specially during the winter.

Raccoon dogs are native to eastern Asia from the Amur River in Siberia to North Vietnam and Japan. They have been introduced into other parts of Siberia and to eastern Europe, where their range is now spreading.

Sleeping dogs?

The native habitat of the raccoon dog is broadleaved forests, usually near rivers, lakes, marshes and rocky gorges. In Europe raccoon dogs have taken to living in coniferous forests. They are nocturnal, lying up by day in dens under rocks, in hollow trees, in dense thickets or in the deserted burrows of foxes and badgers. Some raccoon dogs dig their own burrows with several side passages, and they have even been known to take up residence under occupied houses. When fighting, raccoon dogs whine and growl rather than bark. Each raccoon dog has a home range that varies widely in size, depending on the seasonal food supply.

In the more northerly parts of their range some raccoon dogs appear to hibernate, but it is probably more correct to talk of winter dormancy, as in bears, rather than true hibernation. Local hunters say that only well-fed raccoon dogs sleep for long periods, lying on beds of grass and moss in their burrows. These raccoon dogs emerge on warm days, but those that have not accumulated a good store of fat spend the winter hunting, specially for carrion, and only sleep through blizzards and cold spells.

Eat whatever they find

The diet of raccoon dogs is very varied. They eat plant foods, such as fruits, roots, acorns and grain. A wide variety of animals also features in their diet, particularly insects and small rodents such as voles and gerbils. Other small mammals occasionally taken include hedgehogs, shrews and moles. Raccoon dogs also feed on amphibians, birds (especially ground-nesters such as larks and pheasants), fish, reptiles, mollusks and carrion. Grass snakes are often eaten, and in the southern parts of their introduced range raccoon

dogs eat young tortoises. They also scavenge among garbage. Raccoon dogs have poorly developed carnassial teeth (the dentition typical of carnivores) because much of their diet does not require them.

The raccoon dogs' diet changes according to the time of year. They concentrate on rodents in the spring and take cockchafers and dung beetles as they begin to appear toward the end of spring. Birds and their eggs are eaten in summer and berries and nuts in the fall.

Scent-marked territories

The breeding season is from January to April, when the male's "yearning call" can be heard. A pair of raccoon dogs defends a joint territory, both animals marking it with urine. Gestation lasts 59–65 days, and a litter, usually consisting of 4 to 10 young, is born in the burrow. At first the young are blind and almost helpless. In 3 weeks their teeth erupt, and 1 week later they

In winter raccoon dogs have long thick fur to protect them from the severe cold. They may double their weight to 26 pounds (12 kg) to help them survive winter dormancy.

Foraging on the forest floor. Raccoon dogs were introduced to northern Europe and are now more common there than in much of their native range.

can eat meat brought by the parents, although they still take their mother's milk. The young become independent in the fall, but some families stay together during winter. Raccoon dogs share dens if suitable sites are hard to find.

New European mammal

The raccoon dog has a valuable pelt and is now rare in Japan because it is hunted for both its skin and its flesh. It has been bred in captivity, however. Raccoon dogs have been liberated in parts of Siberia and European Russia as a source of pelts since 1927. They have not done so well in Siberia, where it may be too cold for them, but they have thrived in Russia, especially around Moscow, St. Petersburg, Kalinin and Smolensk and in the Pripet Marshes on the western border. From Russia, raccoon dogs have spread westward. They are now found over most of Poland, as well as in Hungary, Romania and the Czech Republic, and they are spreading into western Germany. They are also found in Scandinavia. Raccoon dogs have spread along the routes of rivers and will probably continue through western Europe along the Oder and Elbe Rivers.

The introduction of a carnivore is usually detrimental to existing wildlife, as proved to be the case with the American mink, *Mustela vison*, in Britain and the red fox in Australia. This is because the introduced mammals often affect the numbers of native animals. However, the raccoon dog does not seem to have become a serious pest in its new homeland.

RACCOON DOG

CLASS	**Mammalia**
ORDER	**Carnivora**
FAMILY	**Canidae**
GENUS AND SPECIES	***Nyctereutes procyonoides***

ALTERNATIVE NAME
Ussuri dog

WEIGHT
Winter: up to 22–26 lb. (10–12 kg); rest of year: 9–13 lb. (4–6 kg)

LENGTH
Head and body: 20–26 in. (50–65 cm); tail: 5–10 in. (12–25 cm)

DISTINCTIVE FEATURES
Stocky body; short legs; black mask around eyes and muzzle; gray or creamy brown fur; dark guard hairs on top of head, along limbs and on tail

DIET
Fruits, roots, nuts, invertebrates, small vertebrates, bird eggs, carrion and garbage

BREEDING
Age at first breeding: 9–12 months; breeding season: January–April; number of young: usually 4 to 10; gestation period: 59–65 days; breeding interval: 1 year

LIFE SPAN
Up to 8 years

HABITAT
Broadleaved forests with good understory, usually near water

DISTRIBUTION
Far eastern Asia, in Russia, China and Japan; introduced to parts of Scandinavia and northeastern Europe

STATUS
Uncommon; rare in Japan

Raccoon dog (native range)

RACER

RACERS ARE CLOSELY related to whip snakes. They have slender streamlined bodies with relatively large heads, small curved teeth and large eyes. All racers are paler on the belly, but depending on the species, of which there are more than 20, they can be black, olive, brown or reddish on the back. Generally, racers living in humid areas are darker than those found in drier parts.

The striped racer, *Masticophis lateralis*, has a marked yellow stripe on its side. Black racers, *Coluber constrictor*, the largest species, are occasionally more than 5 feet (1.5 m) in length, but most specimens are only half this size. The females are larger than the males. The young are usually considerably brighter than the adults and are checkered with alternating blotches of pale olive and darker brown.

Racers are not venomous and they do not constrict their prey, despite the scientific name of the black racer. Instead these snakes capture their prey by simply pinning it to the ground with tight coils of their bodies.

Racers are found in most of the United States, in the extreme south of Saskatchewan, British Columbia and Ontario, and in Central America, south to Guatemala. They are also found in southern Europe and North Africa and in parts of northeast Asia.

Prefer open scrubland

Racers are usually found in scrubland, pastures or among crops, but the larger species also live in open woodland. They are not found in the desert regions of Arizona, New Mexico, Colorado and northwest Mexico. Racers live on the ground but climb into bushes or trees, or even swim, to escape danger. They tend to remain "home based" all their lives, often not traveling out of an area of about 25 acres (10 ha). Racers may move locally, however, for instance in order to escape out of a crop at harvest time or to find a suitable place for hibernation.

Racers hibernate in crevices in rocks or in the disused burrows of pocket gophers or other animals. Sometimes they hibernate entwined in

The weight of the black racer's coils is enough to trap its prey, usually a small lizard, frog, rodent or bird.

A racer's appearance varies with species and location. For example, racers from Ohio, Iowa and Minnesota are sometimes called blue racers, while those from the Midwest are often known as eastern yellow-bellied racers.

BLACK RACER

CLASS	**Reptilia**
ORDER	**Squamata**
SUBORDER	**Serpentes**
FAMILY	**Colubridae**
GENUS AND SPECIES	***Coluber constrictor***

ALTERNATIVE NAMES
At least 9 subspecies of the black racer have separate names

LENGTH
Usually no more than 4 ft. (1.2 m)

DISTINCTIVE FEATURES
Slender, streamlined body; large eyes; smooth scales; color varies enormously depending on location

DIET
Wide range of small mammals, birds, reptiles and amphibians; also large insects

BREEDING
Breeding season: late spring; number of eggs: up to 20; hatching period: 8 weeks

LIFE SPAN
Not known

HABITAT
Very varied; mainly scrub, grassland, fields and open woodland

DISTRIBUTION
Southern Canada south to Guatemala

STATUS
Common

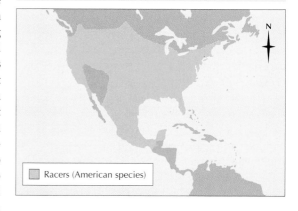

Racers (American species)

---

small groups with other snakes, not necessarily racers. As much as 6 months may be spent in hibernation in the extreme north of their range.

Eating according to size

Racers use eyesight to catch prey, and they often slide through dense cover in active search of food. Smell may also play a part in hunting because they are known to catch some animals in their burrows. When the prey is caught, the racer works its jaws toward the head and swallows the animal whole, head first. The type of food eaten depends largely on the size of the racer. Young snakes feed mainly on grasshoppers, lizards and frogs, but larger snakes will eat mice, voles, birds and other smaller snakes. Adult black racers eat mainly rats, weasels and young rabbits, which they swallow whole. Racers are also said to eat the young of their own species. However, when one naturalist who studied blue racers, *C. constrictor flaviventris*, saw this happen it seemed to have occurred in error. Two racers were trying to swallow the same lizard. One succeeded and proceeded also to swallow the other racer, which was still clinging to the lizard.

Promiscuous snakes

The courtship of racers is similar to that of the garter snakes (discussed elsewhere). The male finds a receptive female in early spring and lies alongside her, rippling spasmodically. The female sometimes moves off, but this seems only to increase the male's interest. Two male racers may court the same female simultaneously. Racers lay eggs in clutches of 10 to 20. The number depends to a large extent on the size of the female. The eggs are white and leathery and may be buried more than 1 foot (30 cm) deep in an abandoned pocket gopher or ground squirrel hole. In humid climates the eggs may be near the surface just under some loose soil or a piece of wood. Racer eggs are unusual because their shells have a knobbly surface; the eggs of most snakes are smooth. Development takes about 2 months and depends on the humidity and warmth of the environment. The young snakes usually hatch in August or September and use an egg tooth to break out of their eggs, which they lose a few days after hatching. The young are about 10 inches (25 cm) long and increase to 17 inches (43 cm) before they hibernate in October.

Will fight if provoked

The natural enemies of racers are hawks, owls, skunks and other snake-eating snakes. The racer's eggs are very vulnerable, not only to predators but also to drying out and flooding. The death rate among the first-year young is very high. Probably about 50 percent die in their first active year, before they have reached mature size. Modern farming has reduced numbers drastically because farm machinery now moves too rapidly to permit escape, deep plowing turns up racer eggs and crop cutting forces many racers to

seek new homes. If disturbed, a racer escapes as quickly as possible, zigzagging silently away in the undergrowth, often using a downhill slope to increase its speed. Sometimes it just coils itself passively and exudes musk or thrashes its tail to attract attention to a certain spot and then glides away quickly, often returning to the same spot by another route. It will also shed its tail, leaving it thrashing on the ground while the rest of the snake escapes. The racer sometimes fights if cornered, raising its head and striking repeatedly at its enemy, lacerating the skin with its teeth.

Racer in name only?

It is commonly accepted that the racer is one of the swiftest of snakes and this is how it got its name. Yet the fastest reliable recorded speed for this same snake is only a little over 3½ miles per hour (5.5 km/h), no more than a brisk walking pace for humans. Compared with the black mamba, *Dendroaspis polylepis*, which attains a maximum speed of 10 miles per hour (16 km/h), the racer turns out to be not at all fast. Even the European grass snake, *Natrix natrix* (discussed elsewhere), travels at 5 miles per hour (8 km/h). The speed of all snakes is probably exaggerated because of the ease with which they slide through undergrowth, disappearing in a trice.

Active by day, racers use their acute vision to catch prey. This young Antiguan racer, Alsophis antiguae, has caught a lizard and is devouring it whole.

RAGWORM

RAGWORMS LIVE MAINLY ON the shore or in shallow seas throughout the world. They are marine relatives of earthworms but have more bristles on each segment and, together with many other kinds of marine ringed worms, are known as bristleworms, or polychaetes (class Polychaeta). Ragworms are active, often bulky, and well-known to sea fishers, who use them as bait, in particular the king rag, *Nereis virens*, which is farmed in Britain to supply the bait market. Ragworms range in size from 1 inch (2.5 cm) long to almost 20 inches (50 cm) long in the case of *Nereis longissima*. They can be shades of brown, red, green or yellow, and their bodies are divided into segments, which may number 200 or more in the longest species. The foremost segment has two short, light-sensitive tentacles and two sensory palps, as well as four eyes. The second segment has a group of four tentacle-like cirri on each side. Each of the remaining segments is flanked by a pair of parapodia, which are flat, lobed outgrowths bearing bristles.

Brain unnecessary for swimming

Ragworms swim by means of side-to-side undulations of the body. These undulations pass forward along the body, instead of backward as in most serpentine animals, thereby allowing the parapodia to act as paddles, pushing the worms forward. Although the brain controls whether movement takes place, ragworms carry on swimming if the head segments are cut off. The worms also continue swimming if they are cut into short lengths, with each length performing the swimming actions. These characteristics demonstrate that it is not the brain that coordinates swimming movements in ragworms but the nerve cord, which runs along the body and thickens in each segment to form a compact mass, or ganglion.

Ragworms spend most of their lives in U-shaped tubes, or burrows, in mud or sand or under stones between tidemarks. Gentle up and down undulations, in contrast to the side-to-side swimming movements, serve to draw currents of water through the burrows. One species, *Nereis fucata*, is noted for its habit of sometimes living in the shells occupied by hermit crabs.

Deadly net

Most ragworms are carnivorous, eating small invertebrates such as crustaceans. Others browse on mud or filter small particles from the water they draw into their tube. *Hediste diversicolor* sometimes catches its food by spinning a net of mucus over the mouth of its tube and drawing water in. Particles of food become caught in the net, and after about 10 minutes the worm pushes its head up to gather the net in its mouth and swallow it. The mouth has a proboscis that turns inside out as it is protruded. At the end of the proboscis is a pair of sharp horny jaws, like those of a beetle. On the sides of the proboscis are groups of other much smaller teeth, or denticles, which help to secure the prey.

In many ragworms breeding involves such great changes in the adults that the breeding worms were once believed to belong to a separate genus, *Heteronereis*. This name survives as a term for the breeding form of a ragworm, although this is now more often referred to as an epitotle. In an epitotle, the hind end of the body becomes swollen with eggs or sperm, and the parapodia become more elaborate and develop fans of long, oar-shaped bristles, more suitable for swimming. The eyes, especially in males, usually become larger, and the tentacles and palps become smaller. The change to the epitotle form occurs only at certain times of the year and is controlled by the brain. If its brain is removed, a young worm will turn into a epitotle immediately. Moreover the epitotle stage can be suppressed in an adult worm by implanting the brain of a young worm.

The epitotles leave their burrows, generally at night, and swarm near the surface, where they release their sex cells. Sometimes they do this in a nuptial dance in which the males swim rapidly in small circles around the females. The eggs are

A European ragworm, Hediste diversicolor, emerges from its burrow. In North America ragworms are also known as mussel worms, clam worms and pile worms.

EUROPEAN RAGWORM

PHYLUM	**Annelida**
CLASS	**Polychaeta**
ORDER	**Phyllodocida**
FAMILY	**Nereidae**
GENUS AND SPECIES	***Hediste diversicolor***

LENGTH
2–5 in. (5–13 cm)

DISTINCTIVE FEATURES
Green, yellow, orange or red, with dorsal blood vessel showing as distinct red line; all Nereidae species easily confused with numerous other polychaete groups and distinguished by relatively large, strong jaws

DIET
Small animals caught with jaws, plus small animals and detritus caught in mucus nets

BREEDING
Sexes separate; breeding season: spring; hatching period: about 1 week

LIFE SPAN
Probably 1–3 years

HABITAT
Usually in black muddy sand of estuaries; also in intertidal and shallow subtidal zones and in offshore waters

DISTRIBUTION
Scandinavia south to northwestern Africa

STATUS
Abundant

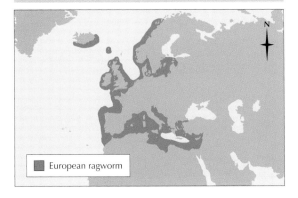

European ragworm

has disintegrated, so there is no obstacle to the sperms reaching the ovary. In some species mating occurs in the tube and the female dies soon after she has laid the eggs. The male then eats her and tends the eggs himself. Ragworm larvae are initially small and spherical. They propel themselves with cilia and bear bristles grouped in three pairs, representing the parapodia of three segments. The number of segments gradually increases and the larvae eventually settle on the bottom.

Survival value of worms

Hediste diversicolor is a ragworm that does not pass through the epitotle form. It lives in the sand and mud of estuaries and shores and can survive in brackish water or even fresh water. It spawns in spring as the temperature of the water rises above 41° F (5° C), and at this time tangles of several females around single males are sometimes to be seen. Both sexes are very fragile and likely to burst if handled. The eggs hatch a little over a week after being fertilized, and the larvae, without bristles at first, stay on the mud surface or in the parental burrow. Having depended on their yolk stores for the first few weeks, they start to feed when they are about 2 millimeters long and make their first burrows when twice that size.

The spread upstream of *H. diversicolor* is limited by the fact that ragworm larvae generally are less able to survive in water of low salinity than are the adults. However, in the freshwater Lake Merced in California lives *Nereis limnicola*, which is closely related to *H. diversicolor*. Possibly both represent one species. The worm survives there because the eggs are fertilized inside the mother, and the larvae emerge only when able to tolerate the fresh water.

In some places digging for ragworms for fishing bait can lead to reduced densities of the worms and disturbance to feeding birds.

released by the rupture of the female's body wall, an event which is stimulated by the presence of the sperm in the water. In the American *Platynereis megalops* fertilization is internal, the male wrapping himself around the female and introducing his sperm into her mouth. At this stage in the lives of both sexes, the wall of the gut

RAIL

THERE ARE ABOUT 130 MEMBERS of the rail family, some of which are treated under separate headings elsewhere in this encyclopedia, including crake, moorhen, coot and gallinule. It is not always easy to distinguish between different groups of rails, but this article deals with those known commonly as rails and that belong mainly to the genus *Rallus*. Rails have short tails and long legs as crakes do, but are distinguished from them by having longer, pointed bills. In some places rails are considered game birds.

The typical rail of Europe and Asia is the water rail, *R. aquaticus*. It is 9–11 inches (23–28 cm) long, a little smaller than a moorhen, from which it is easily distinguished by its long, straight bill and slate-gray breast, neck and face. The upperparts are brown streaked with black, and the flanks are barred black and white. The underside of the tail is white. The water rail breeds in most of Europe, Iceland, parts of North Africa and many parts of Asia east to Japan.

Slightly smaller than the water rail is the 10-inch (25-cm) long Virginia rail, *R. limicola*, which is reddish brown with black-and-white barring on the belly and gray cheeks. It breeds in southern Canada and the northern half of the United States. The clapper rail, *R. longirostris*, and the king rail, *R. elegans*, are nearly twice the size of the Virginia rail and lighter in color. The clapper rail is gray around the head and has a slightly curved bill. It is confined to the eastern and western coasts of the United States and Baja California. The king rail is found in the eastern half of the United States. The African rail, *R. caerulescens*, ranges over most of the southern part of Africa. It is very similar to the water rail but with uniform dark brown upperparts. Similar but smaller is the Lewin's rail, *R. pectoralis*, of Australia and New Guinea, with a chestnut supercilium (the area directly above the eye) and nape and a streaked back. Rails are also found in South America and Madagascar, and many have colonized remote islands.

Wetland birds

Rails are secretive birds that usually live in marshes. The clapper rail is confined to salt marshes along coasts, where the king rail is also sometimes found. Rails' bodies are very narrow, an adaptation for running through dense vegetation such as reed beds. Their unwebbed toes are long, enabling the rails to support themselves on floating plants. Their flight is weak, and rails,

Rails are mostly ground-dwelling water and swamp birds, adapted to living in dense vegetation. The Sora rail, Porzana carolina *(below), ranges across much of North America.*

WATER RAIL

CLASS	**Aves**
ORDER	**Gruiformes**
FAMILY	**Rallidae**
GENUS AND SPECIES	***Rallus aquaticus***

WEIGHT
**Male: 5 oz. (140 g);
female: 3⁷⁄₁₀ oz. (105 g)**

LENGTH
Head to tail: 9–11 in. (23–28 cm)

DISTINCTIVE FEATURES
**Medium-sized bird; very narrow body;
dark brown upperparts with darker streaks;
slate-blue underparts; white-barred black
flanks; long, reddish bill and legs;
unwebbed toes**

DIET
**Wide range of plant matter and small
aquatic animals**

BREEDING
**Age at first breeding: 1 year; breeding
season: eggs laid late March to late July;
number of eggs: 6 to 11; incubation period:
19–22 days; fledging period: 20–30 days;
breeding interval: 1 or 2 broods per year**

LIFE SPAN
Not known

HABITAT
**Freshwater wetlands with flat, muddy areas
and tall aquatic vegetation; often in dense
reed beds**

DISTRIBUTION
**Much of lowland Europe east across Central
Asia and southern Siberia to Pacific, south to
eastern Arabia, northwestern Himalayas and
western China**

STATUS
Locally common

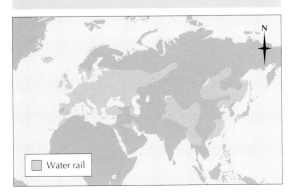

Water rail

like their close relatives, prefer to run to cover when disturbed. When they take to the air, they fly slowly, with their legs dangling. Despite their weak flight, many rails migrate. However, they often come to grief by flying into buildings or overhead wires.

Rails feed on a variety of plants and animals. The water rail includes in its diet watercress, grain, berries, crustaceans, mollusks, worms, insects and sometimes small fish.

Heard but not seen

Many rails have loud and distinctive calls, often heard by night and sounding very eerie. One of the calls of the water rail, known as sharming, sounds like a pig squealing.

Rails make their nests in dense reed beds or aquatic vegetation in swamps or near the banks of rivers and lakes. A rail nest is a bulky but well-concealed mass of dead reeds or sedges. The female rail lays 6 to 11 eggs and both sexes share the incubation for about 3 weeks. The chicks are usually brooded by one parent while the other forages for food, but the blackish chicks soon leave the nest, staying with the parents for about 7–8 weeks.

Although the clapper rail (above) is well represented in North America and has a substantial East Coast population, three West Coast subspecies are endangered, with populations of fewer than 1,000.

Rails are adapted for running through dense waterside vegetation. Although they have unwebbed toes, some species (water rail, above) are able to swim for short distances.

Island rails

Although as a family the rails are poor fliers, they have colonized a number of islands, some very isolated, including Hawaii and Tristan da Cunha. Once on an island, many of these rails evolve to lose their power of flight completely, whereas some fly only feebly. The flightless rail, *Atlantisia rogersi*, of Inaccessible Island, near Tristan da Cunha, has degenerate, hairlike plumage. It is the smallest flightless bird in the world. On these remote islands the rails were once safe. However, most have become vulnerable since the arrival of humans. About 8,400 birds live on Tristan da Cunha. The species formerly known as the Tristan da Cunha moorhen is now known as Gough's moorhen, *Gallinula comeri*. There may be up to 3,000 on Gough Island and about 250 pairs on Tristan da Cunha, but there is a constant risk from mammalian predators that are becoming established on the islands. Other rails are known only from one or two specimens, while four species are known only from bones found in a cave in Bermuda. They were flightless and had died out before humans arrived in the region.

Extinctions and a survivor

One rail species that became extinct relatively recently was the Wake Island rail, *Gallirallus wakensis*. Scientists believe that it was wiped out by a beleaguered and starving Japanese garrison during World War II, who took to eating the bird after they became stranded on the island without supplies.

Another rail species, the Laysan rail, *Porzana palmeri,* also became extinct in the 1940s. This 6-inch (15-cm) long rail was deprived of its native island, part of the Hawaiian chain of islands, in the early 1920s when entrepreneurs introduced rabbits, hoping to start a meat cannery there. The rabbits rapidly devoured the rail's grassy habitat, jeapordizing the bird's existence, although conservationists were able to relocate a few rails to Eastern Island in the Midway Atoll, where they flourished. However, in 1943 a U.S. Navy landing ship ran aground on the island, and rats that had been living on board escaped and began to feed on the Laysan rails. The species was reported to be extinct in 1944. Ironically the rabbits on the Laysan rail's native island were exterminated before the war when the business enterprise failed, and the vegetation gradually regrew.

One island rail, the chicken-sized weka, *Gallirallus australis*, of New Zealand, flourishes despite such onslaughts. Indeed, the bird was once so common that it was widely hunted for its oil. It is flightless but is in no danger of becoming a victim of introduced rodents. A predator itself, it now feeds on rats and mice.

RAIN FOREST

RAIN FOREST MAY GROW wherever a long, sufficiently warm growing season occurs in combination with very high humidity. Thus some temperate regions, such as the Pacific coast of North America or the island of Tasmania, have the right growing conditions for rain forest. The greatest expanses of rain forest are, however, found at tropical latitudes, where radiated solar energy received by Earth's surface is twice that received at a latitude of 60° (in somewhere like Anchorage, Alaska, for example). In those equatorial regions where weather systems regularly roll in from the sea bringing rain, the water combines with this great energy from the sun to produce the richest, most productive and diverse biome on Earth.

The climatic conditions in a tropical rain forest are hot and humid. Daytime temperatures are typically 90° F (32° C), and the annual rainfall is more than 70 inches (1,800 mm). In an equatorial rain forest there is no seasonal variation in day length or temperature. In some regions there is scarcely any seasonality in rainfall either.

Some areas of rain forest are vast. The Amazonian rain forest, for example, is estimated to contain 25 percent of the world's trees. In most regions, however, primary (undisturbed) rain forests have dwindled to small vestiges of their former size. This is due to deforestation and to fires associated with the activities of humans.

Diversity of species

The superabundance of energy and water in the humid Tropics creates ideal conditions for the development of all forms of life. The result is that the organisms that live here tend not to struggle for existence against harsh physical conditions but rather compete intensely with one another. Millions of years of natural selection dominated by the forces of competition have led to a fine differentiation of ecological niches and speciation (the process of species formation). This has resulted in great diversity in all groups of animals and plants, from bats to butterflies. Authors of a US Academy of Science report estimated that 4 square miles (1,000 ha) of primary

The Amazon rain forest is thought to contain 25 percent of all the world's trees. Pictured is the Igapo flooded rain forest in the basin of the Negro River, Amazonas State, Brazil.

Equatorial rain forests are home to a huge diversity of insects, including some of the world's largest and most colorful butterfly species. This is an Australian birdwing, Nithoptera priamus.

RAIN FOREST

ALTERNATIVE NAMES
Selva (tropical rain forest); jungle (term approximates to monsoon rain forest); cloud forest (mountainous rain forest)

CLIMATE
Equatorial rain forest: hot and humid all year with no marked seasonality of temperature and rainfall; average annual rainfall: more than 70 in. (1,800 mm); average temperature: typically 86° F (30° C) by day to 68° F (20° C) at night. Subtropical, monsoon and temperate rain forest: high annual rainfall with distinct wet and dry seasons.

VEGETATION
Tall, fast-growing trees with buttress or stilt roots; epiphytes (plants that derive nutrients from air and rain, and usually grow on trees) and parasitic plants

LOCATION
Equatorial rain forest: Amazon Basin, South America; Congo Basin, central Africa; Sumatra, Indonesia; many Pacific islands. Subtropical rain forest: eastern Brazil; Caribbean and Central America; northern Queensland, Australia; Vietnam; Philippines; much of Indonesia; Madagascar. Monsoon rain forest: Assam and some other states of India; Sri Lanka; Southeast Asia. Montane rain forest: Cameroon, Uganda, Kenya, Rwanda and Burundi; New Guinea highlands; southeastern Venezuela. Mangrove rain forest: low-lying coasts and estuaries in tropical South America, Africa, Indonesia and Australia. Temperate rain forest: British Columbia, Canada; Alaska and Washington, U.S.; southeastern Australia; Tasmania; South Island, New Zealand.

STATUS
Primary (undisturbed) rain forest much reduced in most parts of world, having been destroyed at rapid rate for development

equatorial rain forest can be home to 1,500 species of flowering plants (including 750 tree species), 400 species of birds, 125 species of mammals, 100 species of reptiles, 60 species of amphibians and 150 species of butterflies. Only a tiny proportion of rain forest species has been discovered and scientifically described. In beetles alone, some experts estimate that well over 1 million rain forest species remain unknown to Western science.

The stable climate and lack of seasonality of an equatorial rain forest also contribute to species diversity because animals and plants can reproduce all year, without being restricted to a particular breeding season. This allows, for instance, specialization in fruit eating, as shown by the flying foxes, genus *Pteropus*, of the Old World Tropics. Fruits grow and ripen throughout the year rather than all in one season.

It is often said that equatorial rain forest is very ancient, and that it has existed ever since the Earth recovered from the mass extinction at the end of the age of dinosaurs. Throughout the intervening 65 million years, rain forests have waxed and waned with the passing ice ages. While it is doubtful that any one region has been continuously forested during this period, it is

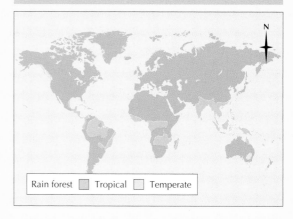

Rain forest [ ] Tropical [ ] Temperate

tempting to speculate that the great age of the forests has contributed to the astonishing diversity and richness of flora and fauna found there.

Massive trees with shallow roots

The architecture of the rain forest is created by the massive rain forest trees, some of which grow to over 200 feet (60 m) in height. Although these trees belong to many different families, for example the Dipterocarpaceae of Indonesian Borneo, they share several superficial similarities. Their leaves are thick and leathery, and often have a drip-tip from which they shed excess water from their surfaces. They lose their leaves only briefly, and at different times from neighboring species, so that at any one time, the forest appears to be fully in leaf. The roots of these trees are very shallow and do not provide sideways support to the tree by anchorage in the soil. Instead, the roots form either buttress or stilt structures that prevent the tree from toppling.

Rising above the rest

The dense growth of trees creates a truly three-dimensional environment, with several vertically stratified (layered) habitats containing specialized communities of animals and plants. The emergent layer is formed by the very tallest trees, the ones that rise above the surrounding forest. This is a discontinuous layer, because emergent trees appear above the canopy only here and there, looking like giant, dark-green cauliflowers. Up here it is warmer, drier and windier than below the canopy, but this is home to a number of rain forest birds and other animals. The monkey-eating eagle, *Pithecophaga jeffreyi*, of the Philippines, for example, often finds the emergent layer a good place to perch and nest.

Life in the canopy

Most rain forest trees grow to 100–150 feet (30–45 m) in height and form a continuous canopy. This is the richest layer of the rain forest and is home to more animals and plants than any other. Birds, bats and insects abound here. Even large, nonflying animals have adapted to move freely around the canopy by a variety of climbing, jumping, gliding and swinging adaptations. In this way, gibbons, sloths and flying lemurs can remain in the canopy, seldom visiting the ground. Small plants can live here too, by adopting an epiphytic lifestyle. Rather than rooting

themselves in the soil far below, epiphytes, which include many orchids and ferns, perch high up on the branches of trees, taking all the moisture and nutrients they need immediately from rain water and from the surface of the tree. Animals have also broken their reliance on the ground for drinking water. They can usually find pools of water in rotten holes in trees and in the pools formed by epiphytic bromeliads (members of the pineapple family). Therefore, most of this wildlife remains unseen from the ground.

Competing for light

Although the forest canopy intercepts 75 percent of the sunlight falling upon the forest, a dense growth of vegetation occurs immediately below, in the understory. Smaller trees are found here, mainly young trees fighting to reach the brightness of the canopy. More epiphytes live here too, and lianas (woody vines) often stretch across any gaps in the vegetation, draping themselves over

A rain forest provides a multilayered habitat. The emergent layer, the canopy and the understory can all be seen in this photograph of broadleaved evergreen forest in Doi National Park, Thailand.

A black-headed parrot, Pionites melanocephala, in the Amazon rain forest, Brazil. Most brightly colored rain forest birds tend to live in the canopy.

the limbs of trees. Many birds make their homes in the understory, but they tend to be less brightly colored than birds living in the canopy. They generally signal to one another by song rather than by striking coloration.

Warm and dark forest floor

By the time sunlight reaches the floor of a primary rain forest, it has only 1 percent of the intensity of the light above the canopy. It is difficult for any but the most specialized plants to live in these conditions. Contrary to the popular, perhaps old-fashioned, image of the jungle, there is very little undergrowth. It is possible for two humans to observe each other when separated by 50 yards (45 m). Specialized plants like the parasitic stinking corpse lilies, genus *Rafflesia*, can live here because they rely on the photosynthetic products of their hosts.

There is little food for herbivorous animals on the forest floor, but some trees produce large fruits that are sent crashing to the ground to be dispersed by the variety of rodents and other animals that live here. Such ground-living mammals include agoutis, tapirs, deer and a host of mice, rats and shrews. Much of the organic waste from life in the canopy also ends up on the forest floor, and it is this that provides the largest resource to be exploited. The forest floor is the most constantly warm and humid layer in the rain forest and temperatures at night may only drop a few degrees below the daily maximum.

These conditions are ideal for decomposition, and a community of detritivores and saprophytes (animals and plants that live off dead or decaying material), fungi and microorganisms feed on the organic waste. They break the waste down at an amazing rate. Among the most numerous of the detritivores are the termites and ants, which comprise an astonishing 40 percent of the total rain forest biomass. However, it should be remembered that most of the mass of trees is dead, unproductive "necromass." Ants inhabit every other habitat in the rain forest as well as the forest floor, taking on a number of different roles in the food web. Some species enter into symbiotic (mutually beneficial) relationships with specialized vegetation, called ant plants, such as the epiphyte *Myrmecodia*.

Temporary clearings

Sometimes, light floods the forest floor in clearings created by fallen trees, and in these cases, conditions quickly become very different. A dense

tangle of undergrowth rapidly develops, as opportunistic plants grab their chance. Young trees emerge as the tallest of these, and soon they are competing fiercely to reach the gap in the canopy. Within 5 years of the gap appearing, seedlings of trees such as *Eucalyptus deglupta* can reach 80 feet (24 m) in height, with a trunk diameter of 1 foot (30 cm). Large, ground-dwelling herbivores such as the African elephant, *Loxodonta africana*, can delay this regeneration process by the destruction of young trees. In so doing, they maintain clearings, and hence the undergrowth on which they can feed.

Rapid nutrient cycle

Decomposition continues beneath the thin surface covering of leaf litter. Specialized burrowing species live here, such as caecilians (legless amphibians), worm lizards and worm snakes.

As soon as decomposition makes nutrients available in mineral form, they are quickly taken up by the trees before they are leached away by the heavy rains. The paradox about rain forest soil is that although the nutrient turnover by decomposition and nutrient uptake is very rapid, this means that there is no time for rich, organic humus to build up. In effect, there is no nutrient surplus in the soil. Once the nutrient cycle is broken, by deforestation for example, the soil quickly becomes infertile, and it is then easily leached and eroded.

Other types of rain forest

The description above is typical of undisturbed equatorial rain forest such as that in the Amazon Basin in South America and in the Congo Basin in central Africa. Not all rain forest is like this, however. For example, disturbance frequently occurs, either by natural phenomena such as landslides and the eruption of volcanoes, or by fires and deforestation by human beings. Areas of disturbed, regenerating forest are called secondary forest, or secondary growth. Secondary forest features more undergrowth, smaller trees and a less continuous canopy than primary forest. These features are also found in what many authorities call subtropical rain forest. This type of rain forest occurs in wet, tropical regions where there is seasonality in rainfall, so that there is a dry season. An extremely marked dry season followed by a very wet rainy season results in what is known as monsoon rain forest, in which many trees are deciduous and there is even denser undergrowth. Subtropical and

The three-toed sloth, Bradypus infuscatus, is one of several large, nonflying animals that spend nearly all their lives up in the canopy. Sloths come down only to visit their communal dung heaps on the forest floor.

monsoon rain forest exhibit greater instability in climate and a pronounced fruiting season. They also support fewer plant and animal species than primary equatorial rain forest.

Further varieties of tropical rain forest occur at high altitudes, in the form of montane forest, and around low-lying coasts and estuaries, where mangrove develops. Lastly, the term rain forest is also applied to a type of evergreen forest found at temperate latitudes. This temperate rain forest also grows under conditions of high annual rainfall, for example the 140 inches (3,550 mm) or more falling annually in the Olympic Mountains, Washington. Here are found forests dominated by the Douglas fir, Sitka spruce and western red cedar, some of which grow to heights of nearly 300 feet (90 m).

Conservation

Tropical rain forests are fast disappearing across all of their former range because the resources they contain are badly needed by the developing economies of the countries in which the forests grow. Forests are destroyed for the purposes of logging, for agricultural land, for the harvesting of fruits, nuts and precious tropical hardwood and for the exploitation of mineral resources. Much of this development is unsustainable, meaning that it cannot carry on indefinitely.

Particularly disastrous is the large-scale clearance of forest for agricultural land. The poor quality of the soil on land recently occupied by rain forest, and the leaching and erosion of the exposed soil, make the land unsuitable for agriculture just a few years after being deforested. Yet more forest clearance is then needed if farming is to continue.

Although tropical rain forest covers 7 percent of Earth's land surface today, some experts estimate that it covered 14 percent 5,000 years ago. Estimates of the present rate of rain forest destruction vary greatly. The Worldwide Fund for Nature (WWF) estimates that 25–50 acres (10–20 ha) are destroyed by logging and fire every minute. This is about 38,600 square miles (100,000 sq km) per year. In the Amazon alone, 6,500 square miles (17,000 sq km) of forest were destroyed in 1999, an area roughly the size of Massachusetts. The Brazilian government has acknowledged that up to 80 percent of logging is conducted illegally.

Tropical rain forests are the product of great climatic stability, so they are vulnerable to any disruption of that stability. While we can see the tragedy of the loss of so much beauty and biodiversity, the short-term livelihoods of millions of people in developing countries continue to depend on the exploitation of tropical rain forest.

An amethystine or scrub python, Morelia amethistina, *resting between the buttress roots of an enormous tree in the rain forest of northeastern Queensland, Australia.*

RAT KANGAROO

R AT KANGAROOS ARE SMALL marsupials that belong to the same order (Diprotodontia) as kangaroos, wallabies, cuscuses, koalas wombats and possums. They have a superficial resemblance to the brown rat, *Rattus norvegicus* (order Rodentia), except that their hind legs are noticeably longer than their forelegs and the tail is usually much shorter than the rest of the animal. There are 10 species, the largest of which is the rufous rat kangaroo, *Aepyprymnus rufescens*. It can be almost 3 feet (90 cm) long, of which nearly 16 inches (41 cm) is tail, and it weighs up to 7¾ pounds (3.5 kg). The smallest is the musky rat kangaroo, *Hypsiprymnodon moschatus*. It can be almost 11 inches (28 cm) long, 4¾ inches (12 cm) being tail, and weighs just less than 1 pound (450 g). A rat kangaroo's dense fur may be harsh or sleek. The muzzle is sometimes naked, and the ears are generally small and rounded. The tail is usually nearly naked, never strongly furred, and is sometimes slightly prehensile. The fur ranges in color from buff to brown and is slightly reddish in some species, being paler, sometimes white, on the underparts. The desert rat kangaroo, *Caloprymnus campestris*, is yellowish ocher with white underparts, the upperparts matching the color of the sand, rock and clay of its habitat in Australia's Lake Eyre Basin.

Rabbitlike warrens

The habitats of the rat kangaroos vary from deserts to the dense rain forests of Queensland, but mostly they live in forests or on grassy plains with scrub or scattered trees. Usually solitary and nocturnal, they rest by day in nests made of grass and other vegetation, sheltered under a tussock or in hollow logs or cavities among rocks. Those species with semiprehensile tails can carry nest materials by wrapping the tail around a bundle. The rufous rat kangaroo is said to make a nest, called a form, in the grass, as does the brown hare, *Lepus europaeus*. One species of short-nosed rat kangaroos, the boodie, *Bettongia lesueuri*, which tends to live in communities, excavates burrows that are almost warrenlike. This species has been known to move into warrens formerly occupied by rabbits or even to share a warren with rabbits. The other rat kangaroos dig only when searching for food. Although not especially fast, rat kangaroos are agile and can dodge or double back. They hop like kangaroos except for the musky rat kangaroo, which is said to move on all fours. Rat kangaroos are alert animals with sharp eyesight and keen hearing.

Diet differences

Often rat kangaroos are mainly vegetarian, eating foods such as fungi, leaves, roots and tubers. The musky rat kangaroo, on the other hand, eats mainly worms and insects, although it also feeds on berries and roots. As is the case with many other marsupials, rat kangaroos seem to be able to go without water when necessary. Rat kangaroos can cause damage to crops, and farmers have sometimes been compelled to take action against them, usually by putting down poison. One short-nosed rat kangaroo is known to hide and store food, and other species probably also have this habit, which further increases the damage to crops.

From the few details known it seems that many species of rat kangaroos breed year-round. Most carry one young in the pouch, but some, such as the musky rat kangaroo, have two. Rat kangaroos also employ a mechanism called

A musky rat kangaroo forages in undergrowth in the Atherton Tablelands, northern Queensland. This species is so named because both sexes have musk glands.

In common with the larger kangaroos and wallabies, rat kangaroos, like this woylie or brush-tailed bettong, have marsupial pouches in which they nurture their young.

RAT KANGAROOS

CLASS	**Mammalia**
ORDER	**Diprotodontia**
FAMILY	**Potoroidae**
GENUS	***Aepyprymnus, Bettongia, Caloprymnus, Hypsiprymnodon*** and ***Potorous***
SPECIES	**10, including rufous rat kangaroo, *Aepyprymnus rufescens*; boodie, *Bettongia lesueur*; and desert rat kangaroo, *Caloprymnus campestris***

WEIGHT
12¾ oz.–7¾ lb. (360 g–3.5 kg)

LENGTH
Head and body: 6–16½ in. (15–42 cm); tail: 4¾–15¾ in. (12–40 cm)

DISTINCTIVE FEATURES
Small size; short, dense, usually brown or sandy fur; long tail; strong hind limbs; elongated snout; small, rounded ears

DIET
Omnivorous; mainly fungi in some species

BREEDING
Age at first breeding: 1–2 years; breeding season: often all year, rainy season only in some species; number of young: 1 or 2; gestation period: 21–40 days, when not delayed; breeding interval: 6–12 months

LIFE SPAN
Up to 13 years

HABITAT
Mainly forests and woodlands; some species: heaths, plains and grasslands

DISTRIBUTION
Australia, including Tasmania

STATUS
***Potorous longipes* and all *Bettongia* species: endangered; desert rat kangaroo: possibly extinct; only 3 species still common**

Rat kangaroos

embryonic diapause, in which a nursing female carries a dormant embryo, the development of which does not begin until the mother has weaned her current young.

Running out of habitat

Rat kangaroos have many common names, some derived from their aboriginal names. Among the short-nosed rat kangaroos (genus *Bettongia*) are the bettongs, the boody and the woylie (*B. penicillata*), while the long-nosed rat kangaroos (genus *Potorous*) are known as potoroos. When there are many common names, the animals bearing them can usually be said to be plentiful or to have once been so. However, only three of the 10 species of rat kangaroos are still reasonably common. The causes of their decline include predation by introduced animals such as the red fox, *Vulpes vulpes*, hunting by aborigines and poisoning by farmers. However, the biggest threat has come from the reduction in suitable habitat as a result of the building of human settlements and intensification of farming practices. The changes in the frequency and severity of forest fires across Australia may also have played a part.

The desert rat kangaroo and one species of the short-nosed rat kangaroos may already be extinct. The former, first recorded in 1843, was rediscovered in 1933, when 17 were sighted and some were caught. They were reportedly hard to run down because they showed such endurance. One succeeded in exhausting two galloping horses in a 12-mile (19-km) chase and was only brought to a standstill, quite spent, by a third horse. However, the desert rat kangaroo has not been seen since the early 1930s.

RAT SNAKE

In North America the name rat snake is used for a number of species that belong to the genus *Elaphe*. Confusingly, however, not all snakes within this genus are called rat snakes. Other members of the genus in North and Central America include the corn snake, *E. guttata guttata*, and the fox snake, *E. vulpina*. To add to the confusion, two subspecies of corn snakes are called rat snakes.

Other species of *Elaphe* are found in Europe, including the four-lined snake (*E. quatuorlineata*), which is the largest European snake; the leopard snake (*E. situla*); the aesculapian snake (*E. longissima*); and the ladder snake (*E. scalaris*). There are other species in various parts of Asia, such as the Indian rat snake, *Ptyas mucosus*.

Rat snakes derive their name from the fact that rodents form a major part of their diet. In other parts of the world unrelated rodent-eating snakes are also sometimes called rat snakes. A characteristic feature of the North American species is that their belly is very flat. They are sometimes described as having a cross section like a loaf of bread. Many forms of *Elaphe* are brightly colored, but others have a relatively dull and unpatterned appearance.

Many names

Of the North American rat snakes, the species with the most widespread distribution, *E. obsoleta*, is often called simply the rat snake. It is very variable in appearance. Most biologists consider that it can be divided into a number of subspecies. It grows to a total length of 6 feet (1.8 m). The black rat snake, *E. o. obsoleta*, ranges from New England and Ontario to Wisconsin and Oklahoma. On much of the East Coast and in Florida it is replaced by the yellow rat snake, *E. o. quadrivittata*. These snakes are not always yellow in color but they nearly always have four dark stripes running along the length of the body.

Everglades rat snakes, *E. o. rossaleni*, are rather similar but are often orange in color. Gray rat snakes, *E. o. spiloides*, are found from central Georgia to the Mississippi Basin. They are usually marked with prominent brown or gray blotches. Texas rat snakes, *E. o. lindheimeri*, may also be marked with dark blotches, but these are less prominent than those in gray rat snakes. The skin between their scales is sometimes red.

Other species in North America include the green rat snake, *E. triaspis*, found in Arizona and Mexico. Rosy rat snakes, *E. guttata rosacea*, and Great Plains rat snakes, *E. g. emoryi*, are subspecies of the corn snake.

Varied diet

North American rat snakes are constrictors, trapping and strangling their prey in the coils of their bodies. The diet is unusually varied for snakes, including frogs, lizards, other snakes, birds and mammals, although rodents such as rats, mice, voles and squirrels form the bulk of the diet.

Twining courtship

In the aesculapian snake, as in many other species, courtship takes the form of a chase followed by a dance. The male pursues the female until he can coil around her. The two snakes continue in this position, and then rear up and dance for up to an hour or more before

North American rat snakes, such as this juvenile black rat snake, have slightly keeled scales, which make them very agile tree climbers.

Although rat snakes are nonvenomous, they are often killed by people because they are aggressive when threatened and may vibrate their tails like rattlesnakes.

copulation takes place. Some rat snakes are oviparous, laying clutches of about 20 eggs in burrows in loose earth or decaying logs. The eggs are laid in midsummer. The female rat snake sometimes stays with them until they hatch, about 9 weeks later. She then leaves.

Foul defense

A rat snake's first line of defense is to emit a foul-smelling fluid from glands at the base of the tail. Many other snakes do this too, for example the grass snake, *Natrix natrix*, of Europe. Some species vibrate their tails when annoyed or cornered, and the fox snake was once called the hardwood rattler because of this characteristic. However, this habit does not protect the harmless rat snake from humans who generally brand them as dangerous. Indeed, it often leads to the rat snake being confused with the rattlesnake, and being killed on sight. If handled, a rat snake is aggressive and may bite and draw blood, although it is not venomous.

Beneficial snakes

The aesculapian snake is believed to be the snake wrapped around a staff on the symbol of Aesculapius, the Greek god of medicine. The original temple of Aesculapius was at his supposed birthplace at Epidaurus, in Greece. It was frequented by large, easily tamed snakes that have been named aesculapian snakes. They were thought to be incarnations of the god and to have healing powers.

The Romans later took over many of the Greek myths, including that of the aesculapian snake. Furthermore, they took the snake with them on their travels, which accounts for isolated colonies of aesculapian snakes in Europe as far north as Germany.

RAT SNAKES

CLASS **Reptilia**

ORDER **Squamata**

SUBORDER **Serpentes**

FAMILY **Colubridae**

GENUS AND SPECIES **Many North American species, including corn snake, *E. guttata*; rat snake, *E. obsoleta*; fox snake, *E. vulpina*; and green rat snake, *E. triaspis***

ALTERNATIVE NAMES
Subspecies of *E. obsoleta*: black rat snake; mountain black rat snake; yellow rat snake; gray rat snake; Everglades rat snake; Texas rat snake; chicken snake; pilot snake; striped house snake. Subspecies of *E. guttata*: Great Plains rat snake; rosy rat snake.

LENGTH
3½–6 ft. (1–1.8 m)

DISTINCTIVE FEATURES
Strongly flattened belly; slightly keeled scales; color varies depending on location

DIET
Wide variety of small mammals, particularly rats; also birds, eggs, reptiles and amphibians

BREEDING
Breeding season: midsummer; number of eggs: about 20; hatching period: 60–65 days

LIFE SPAN
Not known

HABITAT
Almost all terrestrial habitats except hot deserts and above tree line in mountains

DISTRIBUTION
New England south to Florida, west to Wisconsin and Texas; isolated population in Colorado, Utah and northern Mexico

STATUS
Generally abundant or common

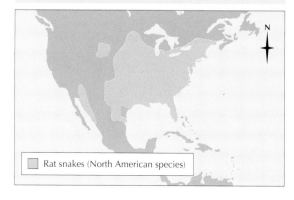

Rat snakes (North American species)

RATTLESNAKE

RATTLESNAKES ARE HEAVY-BODIED and usually highly venomous snakes, best known for the rattle at the tip of the tail, which is also known as the buzzer, whirrer, bell or cloche. When it is disturbed, a rattlesnake vibrates its tail, which causes the sound from the rattle. Rattlesnakes are found only in the New World. There are about 30 species, most of which occur between southern Canada and the south of Mexico, but three species have ranges that extend into South America as far as Argentina.

There are two groups of rattlesnakes, each represented by one genus. Pygmy rattlesnakes (*Sistrurus*) rarely exceed 2 feet (60 cm) in length and have tiny rattles. True rattlesnakes (*Crotalus*) are usually 3½–5 feet (1–1.5 m) in length, although one eastern diamondback, *C. adamanteus*, was measured at 8 feet (2.4 m), the greatest recorded length for a rattlesnake. In common with other pit vipers (described elsewhere in this encyclopedia) rattlesnakes are able to tolerate low temperatures, and the ranges of several species extend to altitudes of more than 11,000 feet (3,350 m).

A rattlesnake's rattle

The rattle is made up of a number of loosely interlocked shells, each of which was originally the scale that covered the tip of the tail. Usually in snakes this scale is a simple hollow cone that is shed with the rest of the skin at each molt. In rattlesnakes it is larger than usual, much thicker, and has one or two constrictions. Except at the first molt, the scale is not shed but remains loosely attached to the new scale, and at each molt a new one is added.

The rattle does not grow in length indefinitely. The end scales tend to wear out, so there can be a different number of segments to the rattle in different individuals of the same age, depending on the extent to which the end of the rattle is abraded. It seldom exceeds 14 segments in wild rattlesnakes, no matter how old they may be, but snakes in zoos, which lead a less hazardous life and do not rub the rattle against hard objects, may have as many as 29 pieces in a rattle. The longer the rattle, the more the sound is deadened, eight segments being the most effective number to produce the loudest noise.

The species C. viridis *is the most widespread rattlesnake in the United States. It is also the most variable, with 9 subspecies, including the prairie rattlesnake,* C. v. viridis *(above).*

A Mohave rattlesnake, C. scutulatus. The rattle probably evolved so the snakes could warn large grazing mammals such as bison of their presence, and avoid being trampled.

The volume of sound varies not only with the size of the snake and the length of the rattle, but also from species to species.

Variation among species

It is as difficult to generalize about the size and effectiveness of the rattle as it is to do so about any other feature of rattlesnakes. For example, these snakes have a reputation for attacking people and being bad-tempered, but this applies only to some species. Unless provoked or roughly treated, the red diamond rattlesnake, *C. ruber*, may make no attempt to strike when handled. It may not even sound its rattle. By contrast, the eastern diamondback, *C. adamanteus*, and the western diamondback, *C. atrox*, not only rattle a warning but also pursue intruders, lunging at them again and again. How poisonous a snake is also depends on several things, such as its age (the younger it is, the less the amount of poison it can inject) and whether it has recently struck at another victim (if it has done so, the amount of venom it can use will be reduced). Snakes have been known to take nearly two months to replenish their venom to full capacity. Rattlesnakes of the same species from one part of the range may be more venomous than those from another part. For example, prairie rattlesnakes, *C. viridis*, that live in the plains are about three times as venomous as those of California and half again as venomous as those of the Grand Canyon.

Waterproof skin

Rattlesnakes feed on much the same prey as other pit vipers, mainly small, warm-blooded animals, in particular rodents, cottontail rabbits

RATTLESNAKES

CLASS **Reptilia**

ORDER **Squamata**

SUBORDER **Serpentes**

FAMILY **Crotalidae**

GENUS AND SPECIES **Eastern diamondback rattlesnake,** *Crotalus adamanteus*; **western diamondback rattlesnake,** *C. atrox*; **timber or banded rattlesnake,** *C. horridus*; **red diamondback rattlesnake,** *C. ruber*; **tiger rattlesnake,** *C. tigris*; **Aruba Island rattlesnake,** *C. unicolor*; **prairie rattlesnake,** *C. viridis viridis*; **Massasauga,** *Sistrurus catenatus*; **pygmy rattlesnake,** *S. miliarius*

ALTERNATIVE NAMES
Rattlers (all species); coontail rattler (*C. atrox*); ground rattler (*S. miliarius*); swamp rattler, black snapper (*S. catenatus*)

LENGTH
Usually 3½–5 ft. (1–1.5 m)

DISTINCTIVE FEATURES
Stout body; large fangs; tail rattle

DIET
Mainly small mammals

BREEDING
Viviparous. Number of eggs: usually 10 to 20; breeding interval: often 2 years.

LIFE SPAN
Not known

HABITAT
Very varied, from forest to arid desert

DISTRIBUTION
Southern Canada south to Venezuela; northeast coastal regions of South America; southern Brazil; northern Argentina

STATUS
Some species locally common; endangered: *C. horridus*; threatened: *C. unicolor*

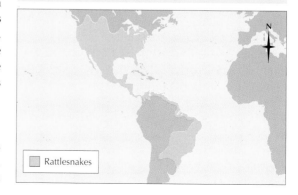

Rattlesnakes

and young jack rabbits. Young rattlesnakes, including the pygmy rattlesnakes, take a larger proportion of cold-blooded animals, such as frogs, salamanders and lizards. Rattlesnakes' water needs are not as great as those of active and warm-blooded animals because the water loss from the body is not high. They need only about one-tenth as much water as a mammal of similar size.

In one test scientists found that twice as much water is lost from a rattlesnake's head, mainly in its breath, as from the whole of the rest of its body, which suggests that its skin is almost waterproof. When it does drink, the snake sucks up water from a pond or stream. There is no evidence that it laps it up with the tongue, as is sometimes stated, or that it drinks dew.

Two years to be born

All rattlesnakes give birth to live young. Whether they have one litter a year or less depends on the climate. The prairie rattlesnake has one litter a year in the southern part of its range, but in the northern part it may be two years before the young are ready to be born. Mating takes place in spring, and the number in a litter may vary from 1 to 60 according to the size of the mother, the usual number being between 10 and 20.

Their venom does not prevent rattlesnakes from being killed and eaten. Hawks of all kinds kill them, as do skunks and snake-eating snakes. Pigs, deer and other hooved animals trample them, especially the young ones, and many die of cold, excessive heat or starvation. Indeed, few rattlesnakes from a litter survive their first year.

A snake's senses

Snakes are known to be deaf, yet they often seem to react to sounds. However, in his experiments on rattlesnakes the American herpetologist Laurence Klauber found that this was not the case. Having placed a red diamond rattlesnake under a table, Klauber first clapped two sticks together, making sure his hands and the sticks could not be seen by the snake. It reacted, apparently to the sound. Klauber was puzzled by the reaction at first. However, he finally worked out the cause. He was sitting on a stool, his feet dangling, and every time he clapped the sticks together his feet moved and the snake reacted to the sight of them. Following this discovery, Klauber placed a screen between the snake and his feet before repeating the test. To his surprise, the snake reacted once again when he clapped the sticks, this time because it could see a reflection of Klauber's feet in a nearby window.

Klauber found that the rattlesnake was highly sensitive to footsteps up to 15 feet (4.6 m) away on a concrete floor. The snake still reacted to footsteps this distance away after Klauber had placed it on a blanket, in an effort to absorb any vibrations that the snake might pick up from the floor. Deciding to test this further, he put the snake in a fiberboard box, suspended this by a rubber band from a stick and held each end on a pillow to insulate it from vibrations through the ground. It still reacted to clapped sticks and to a nearby radio that was switched on. Klauber finally determined that the snake was picking up the heat from the valves of the radio as they warmed up, and it was reacting to vibrations in the floor and sides of the fiberboard box, against which its body rested. The box was changed for a Chinese woven bamboo basket and hung from the same stick. The snake still appeared to react to sound, but further tests showed it was reacting to Klauber's hand movements seen through the very tiny cracks between the bamboo.

Klauber's experiments indicate how hard it sometimes can be to test a particular animal sense. They also show, among other things, how sensitive a snake's eyes are to small movements.

The western diamondback injects venom into its prey through a pair of hollow fangs. Like all snakes, it is able to dislocate its jaws and swallow its prey, usually small mammals, whole.

RAVEN

has sharper and more elongated throat feathers than those of the crow, which are more rounded. The raven's call, known as a caw, is deeper, throatier and less repetitive than the crow's.

Ranging over a variety of climates worldwide, the common raven is found across Europe and in the deserts of northern Africa, in Asia except for the south and southeast, in Greenland and in America as far south as Nicaragua. The raven lives mainly in mountainous and wild hilly areas and on cliffs as well as in lowland woods. The precise choice of habitat varies according to region: it also includes treeless tundra, riverbanks, rocky cliffs, plains and deserts. In fact, the raven seems able to live in most habitats, up to an altitude of 14,700 feet (4,480 m). However, the raven favors wooded areas, especially those situated along the coast and in mountain regions. Cliff ledges or trees are favorite sites for nesting.

The name raven has been given to other large members of the same genus in other parts of the world. These are the Chihuahua raven, *C. cryptoleucus*, of North America; the African white-necked raven, *C. albicollis*, of eastern and southern Africa; the thick-billed raven, *C. crassirostris*, of northeastern Africa; the fan-tailed raven, *C. rhipidurus*, of northeastern Africa and the Middle East; and the Australian raven, *C. coronoides*. Their habits are similar to those of the common raven. Other raven species include the little raven, *C. mellori*, and the forest raven, *C. tasmanicus*, also found in Australia.

The raven is the largest of all birds that have an entirely black plumage. The sexes are very similar in color, but the female may be distinguished by her smaller size.

THE COMMON RAVEN, *Corvus corax*, is the largest member of the crow family, a family regarded by many ornithologists as containing the most intelligent of all birds. The raven is all black, its plumage having a blue-purple iridescence, and grows to about 25 inches (64 cm) long, making it the largest passerine (perching bird). It has a heavy bill and wedge-shaped tail. When flying, with its neck outstretched, its pointed throat feathers stand out, giving a rufflike effect. The raven is often confused with the crow, but a number of features distinguish the two birds. The raven's bill is larger and heavier than the crow's, and the raven

Playful ravens

The raven's flight is powerful, with regular wingbeats, and it is easily identified on the wing not only by its size but also by its deep call of *corronk*. It lives in pairs or small groups but roosts in larger numbers of up to 100 or more. Evidence suggests that the raven mates for life.

Ravens perform a range of aerial displays, although these are not always easy to observe as wild ravens are wary birds, particularly during the nesting season. They glide and soar, but also indulge in aerobatics, nose-diving with closed wings, sometimes gliding upside down for a short distance in a corkscrew roll, somersaulting

COMMON RAVEN

CLASS	**Aves**
ORDER	**Passeriformes**
FAMILY	**Corvidae**
GENUS AND SPECIES	***Corvus corax***

WEIGHT
Male: 38–55 oz. (1,080–1,560 g).
Female: 28¼–46 oz. (800–1,315 g).

LENGTH
Head to tail: about 25 in. (64 cm);
wingspan: 47–55 in. (1.19–1.4 m)

DISTINCTIVE FEATURES
Large size; all-black plumage; powerful bill;
shaggy throat; fairly long, wedge-shaped tail

DIET
Wide-ranging diet includes carrion, insects,
mollusks, small and medium-sized birds,
small mammals, frogs, lizards, eggs, nuts,
berries, grain and garbage

BREEDING
Age at first breeding: 3–4 years; breeding
season: eggs laid mainly January–April;
number of young: 4 to 6; incubation period:
20–21 days; fledging period: 35–42 days;
breeding interval: 1 year

LIFE SPAN
Up to 10 years or more

HABITAT
Mainly mountains, hills, cliffs and woodland;
also treeless tundra, coastlines, riverbanks,
montane forest, plains and deserts; feral
populations in villages, towns and cities

DISTRIBUTION
North and Central America; southern
Greenland; Europe; North Africa; northern
and Central Asia

STATUS
Generally common, but population declines
in many areas due to persecution

Common raven

or turning, twisting and tumbling. These performances form part of the courtship ritual, but they also take place at other times, when they have the appearance of play.

There are certain behavioral acts that ravens indulge in regularly. One raven will pull another's tail. Two will play with their bills as though they were kissing. Ravens will give sticks and stones to each other, or several may pass them from bill to bill, pouncing and grabbing at them or offering a stick to another and jerking it away when the other goes to take it. An open bill indicates threatening behavior.

The raven's powerful bill enables it to consume a wide range of food types, including mollusks, acorns and grain as well as small mammals, birds, reptiles and frogs.

Universal scavenger

The raven's varied diet consists mainly of carrion, and where ravens are numerous, they gather around a large carcass as vultures do. They sometimes kill weakened lambs, small mammals such as rabbits, rats, mice, moles and hedgehogs, small and medium-sized birds, even those as large as ducks, and lizards and frogs. They also take eggs, snails and insects. In fact, a raven scavenges anything edible, in forests, in fields, on the shore and in deserts. Ravens also eat grain, acorns, cherries and beech mast. Like other members of the crow family, they bury surplus food, digging small holes in the ground, putting the food in and covering it with grass and earth, going back to their stores when food is scarce. However, most of this food is never recovered, and the crow family is probably a significant factor in the distribution and natural regeneration of many trees. Eurasian jays doubtless help in the spread of oak trees uphill.

Sturdy nests

Courtship consists of the aerobatics already described, after which the pair land, the male making a musical vibrating call. He caresses his mate's bill with his and tickles her under the chin with his bill, and then the two bring their bills together in a "kiss." Both birds of a pair combine to build a solid nest of sticks high among the rocks or in tall trees. It is lined with moss, leaves, grass, hair or wool. In February or March, in much of the range, four to six light blue to greenish eggs, spotted and blotched with brown, are laid. The hen alone incubates them, being fed by the cock. The eggs hatch in 3 weeks, the nestlings being fed by both parents for 5–6 weeks. The parents roost away from the nest soon after the eggs hatch. This habit may be connected to the adage that ravens are unkind to their young. Indeed, the collective noun for the birds is an "unkindness" of ravens.

Persecuted by humans

The raven is a heavy, powerful bird and can defend itself against a peregrine or a group of gulls. To ward off a falcon, or a bird of a similar size, it may roll over, bringing itself belly-up in midair, as in its courtship and play antics. The greatest threat to the raven comes from humans, who hunt it because of its attacks on lambs and poultry, as well as its threat to game preserves. Today raven numbers are severely reduced in many parts of its range, and it has been eliminated altogether in other places. For example, it was once a common sight in London streets but over time persecution by humans and changes to its habitat drove the bird northward and westward to sea cliffs. Now it is spreading back eastward again from its heartland in the western half of Britain and in the Scottish Highlands.

The raven became extinct in the Netherlands in 1924, but in recent years a small population has become reestablished there. Due both to active conservation and to reduced persecution, this expansion has also been noted in many other European countries.

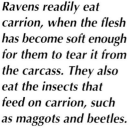

Ravens readily eat carrion, when the flesh has become soft enough for them to tear it from the carcass. They also eat the insects that feed on carrion, such as maggots and beetles.

RAZORBILL

RAZORBILLS, OR RAZORBILLED AUKS, have the same upright, penguinlike stance as murres, also known as guillemots, members of the genus *Uria*. At 14½–15½ inches (37–39 cm) long, razorbills are about the same size as murres, but plumper. Razorbills are named for the sharp edges of their deep, puffinlike bills. The plumage is black and white; in summer the head and neck are black, but in winter white extends from the breast to the throat, chin and sides of the face. The bill is crossed with a white line. The secondary wing feathers are tipped with white, forming a white trailing edge to the wings in flight and a white line across the back at rest.

The range of razorbills is limited to the North Atlantic. They breed on coasts from western Greenland to New England in the west and from Bear Island to Brittany in northwestern France in the east. Much of the year is spent well out to sea, the birds only moving inshore to breed. In the Baltic Sea they may be found on either brackish or fresh water. This is unique among auks, which are nearly all marine seabirds.

Nest on cliffs

Razorbills often breed in the same places as murres, although they are usually less abundant and are not as sociable. They nest on the same inaccessible cliff ledges but are also found among boulders piled up at the bases of cliffs or on rocky shores, where murres are absent. The birds glide down from the cliffs with an unusual slow wingbeat, but travel to and from the feeding grounds in parties, with the auks' typical fast, whirring flight. Auks are difficult to distinguish at a distance when sitting on water, but a razorbill floats higher than a murre and its longer tail is usually tilted upward. Like other auks, razorbills swim underwater with their wings, and they molt all their flight feathers at once so they are flightless for a time during late summer.

The colonies are deserted after nesting and the razorbills move away, spending the winter on coastal waters on both sides of the Atlantic. The northern populations move south, and some spread into the Mediterranean as far as Italy.

Noncompeting fishers

Razorbills eat mainly fish such as sand eels and smelt; in the Mediterranean they have been recorded as eating sardines and anchovies. They also catch planktonic worms and mollusks, and crustaceans such as amphipods. They carry fish crosswise in the bill, in the manner of puffins. Up to eight fish can be carried at one time.

Razorbills, puffins and murres do not seem to compete for food. Murres eat fairly large fish, which they carry singly, lengthwise, in the bill. Puffins carry many small fish in their large bills. Razorbills, with compressed bills that are between the slender bills of murres and parrotlike bills of puffins in size, take medium-sized fish.

Single egg

Razorbills arrive at their colonies from the end of January onward, but breeding starts in earnest in April and eggs are laid in early May or June, depending on latitude. There is no nest except sometimes for a few plants or stones, and the single egg is usually laid in the shelter of a boulder or in a crevice, sometimes inside a puffin or rabbit hole but only infrequently on the exposed ledges chosen by murres. The brood patches, areas of skin devoid of feathers where the eggs are warmed, are small and are found on each side of the breast, so the egg lies under the wing of the brooding parent. As there are two brood patches, it was once erroneously thought that razorbills laid two eggs.

The white line running from the base of the bill to the eye is a feature of the razorbill in mating season. The black throat and neck sides occur only in summer; in winter these are white.

Although it is locally common, the razorbill is vulnerable to threats such as oil pollution, overhunting of its prey and gill-netting. Seventy percent of the global razorbill population is in Iceland.

Incubation takes 35–37 days and the chick is fed by both parents. When they are about 18 days old, those chicks that have survived attack by black-backed gulls launch themselves from the cliff, fluttering down to the sea on their tiny wings. Here they join their parents and swim out to sea, to be fed until they can fly and fend for themselves. The chicks are well protected by fat and feathers and can swim strongly, so they survive falling onto rocks or into heavy surf.

Razorbills and oil

When the tanker *Torrey Canyon* was wrecked off Cornwall in southwestern England in March 1967, its cargo of oil gushed out into the sea and was washed onto the shores of southern England and Brittany. Both areas are rich in seabirds, and it soon became apparent that large numbers were becoming fouled with oil. About 10,000 birds were rescued and attempts were made to clean them. However, most died and it is estimated that as many as 100,000 seabirds perished in all.

Oil pollution has threatened seabirds for many years now, and it was soon clear that it was the auks, especially razorbills and murres, that suffered. A large percentage of razorbills and murres were oiled, probably because they spend more time on the water than gulls or cormorants, which feed at sea and then return to land to roost. Auks are also likely to dive when in trouble and so collect more oil. The wreck of the *Torrey Canyon* dealt severe blows to the auk populations of southern England and Brittany.

RAZORBILL

CLASS	**Aves**
ORDER	**Charadriiformes**
FAMILY	**Alcidae**
GENUS AND SPECIES	***Alca torda***

ALTERNATIVE NAME
Razor-billed auk

WEIGHT
18½–3½ oz. (525–900g)

LENGTH
Head to tail: 14½–15½ in. (37–39 cm); wingspan: 25¼–26 in. (63–66 cm)

DISTINCTIVE FEATURES
Plump body; thick neck; large, thick black bill with white stripe across both mandibles (bill-halves); black webbed feet set far back on body; black upperparts and head; white underparts

DIET
Mainly fish, including sand eels, smelt, sardines and anchovies; planktonic worms; mollusks; some crustaceans

BREEDING
Age at first breeding: 4–5 years; breeding season: eggs laid May–June; number of eggs: 1; incubation period: 35–37 days; fledging period: 17–19 days; breeding interval: annual

LIFE SPAN
20 years or more

HABITAT
Seas and oceans, mainly in offshore waters and along coasts; nests on coastal cliffs

DISTRIBUTION
North Atlantic and Low Arctic, from Greenland and Iceland south to Maine and western Mediterranean

STATUS
Locally common

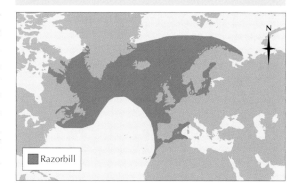

Razorbill

RAZOR SHELL

RAZOR SHELLS ARE MOLLUSKS of the family Solenidae. They have long, narrow, open-ended shells and are admirably suited to burrowing in sand. The shells are usually 5–6 inches (13–15 cm) long and ½ inch (13 mm) or so across, although the pod razor shell, *Ensis siliqua*, can be up to 8 inches (20 cm) across. Their shape, so suggestive of the old-fashioned razor or of a jackknife, has earned them the name of jackknife clam in the United States. The hinge and ligament that join the two straight or slightly curved valves, or shell halves, is forward of the lengthwise midpoint, indicating that the shape of the shell is achieved mainly by the lengthening of its hind end. Close relatives of the true razor shells are the mollusks of the family Pharidae. Among these is the European eggshell razor, *Pharus legumen*. It has a similar elongated shell to the razor shells of the Solenidae, but the hinge is central. Another member of the Pharidae is the transparent razor, *Pharus pellucidus*, which is only 1½ inches (4 cm) long.

Hydraulic feeding

Razor shells live on or close to the shore and feed only when the tide is in, resting just below the surface of the sand with their short siphons projecting. Water, containing oxygen and particles of food, is drawn in through the siphon nearest the hinged side of the shell. As is the case in most bivalves, the food particles are caught up in mucus on cilia-covered gills and propelled from there to a pair of palps on either side of the mouth. The triangular palps lie near the middle of the shell and sort the particles by means of a complex arrangement of cilia, sending some to the mouth and others into rejection currents leading to the second siphon, through which flows a steady outward current of water. Every now and then the adductor muscles pull the two shell valves suddenly together, causing water and waste products to be ejected from this siphon. The siphons are the most exposed parts of a razor shell and bear a crown of pigmented tentacles carrying sense organs.

Spoutfish

When the tide is out, razor shells usually retreat below the surface of the sand, but their presence is sometimes revealed by shallow depressions on the exposed beach. Jets of water and sand are suddenly forced out from these depressions, so giving rise to the mollusk's nickname of "spoutfish." To see the animals producing these jets, one must approach with caution, because they are highly sensitive to vibration and quickly retreat farther down into the sand, perhaps to a depth of 2–3 feet (60–90 cm). A sudden lunge with a spade may produce nothing more than half an animal, often less, and it is easier simply to place a pinch of salt over the hole and wait for the irritated animal to surface.

How they burrow

A razor shell moves up and down in its burrow using its muscular foot at the downward end combined with movements of the shell. At rest, the foot is about half the length of the shell, but it can be shortened still further or extended to about the full length of the shell. The foot can grip sand with such tenacity that it may be pulled off if the shell is tugged hard. The tip of the foot is normally pointed, but in burrowing this becomes swollen into a bulbous disc by blood forced into it from the rest of the body. To dig, the animal contracts is foot muscles, drawing the shell downward. At the same moment the mollusk draws its streamlined valves closer together using the shell adductor muscles, easing the shell away from the walls of the burrow. Then, as the foot is being pushed forward into the sand, the animal allows the

Empty pod razor shells lie on a beach. At 8 inches (20 cm) long, the pod razor is the largest species of razor shell.

A pod razor shell,
Ensis siliqua, *its foot exposed, begins to dig a burrow. These razor shells live on or close to the shore, although close relatives such as the transparent razor,* Pharus pellucidus, *may occur down to 330 feet (100 m) or more.*

RAZOR SHELL

PHYLUM	**Mollusca**
CLASS	**Bivalvia**
ORDER	**Eulamellibranchia**
FAMILY	**Solenidae**
GENUS AND SPECIES	***Ensis ensis***

ALTERNATIVE NAME
Spoutfish

LENGTH
About 5 in. (13 cm)

DISTINCTIVE FEATURES
Long, curved razor-shaped shells

DIET
Filter feeder

BREEDING
Sexes separate; eggs and sperm shed into water; eggs fertilized in water; larvae in plankton for some weeks before settling in sandy areas

LIFE SPAN
Probably up to 20 years

HABITAT
Fine sand on lower shore

DISTRIBUTION
Coasts from Scandinavia south to northwestern Africa; also in Mediterranean

STATUS
Common to abundant

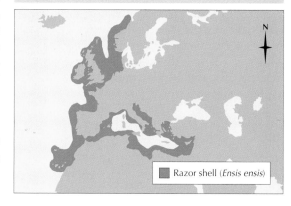

Razor shell (*Ensis ensis*)

valves to press outward to increase their grip. The razor shell uses the same combination of movements to travel upward in the burrow, the foot then pushing the shell toward the surface instead of pulling it down. With the end of the foot anchored, blood is forced into its upper regions, causing the foot to elongate while the shell valves are drawn together.

Swimming razors

Razor shells do not often leave their burrows, but are far from helpless out of sand. They can leap several inches with a flick of their foot and can swim backward with a sort of jerky jet propulsion. In this movement the water enters largely through the siphons and is suddenly expelled through a small hole between the foot and a collar of tissue surrounding it just inside the shell. The water is driven out forcibly by a pumping action in which the shell valves are drawn together and the pistonlike foot is suddenly retracted. Small razor shells can swim so well they are very occasionally taken in tow nets near the sea surface.

Razor slashing

Razor fish (the contents of razor shells) are not eaten much today but were esteemed in the past and are mentioned by various Greek authors, including Aristotle. They were among the delicacies said to have been served at the marriage feast of Hebe, daughter of Zeus. Zoologist John Gwyn Jeffreys, writing about these animals in the mid-nineteenth century in his book *British Conchology*, passed on a curious account of the capture of razor fish at Naples, in southern Italy. There, where the water is shallow, the fisherman sprinkled oil on the surface to see the holes in the sand more clearly. "He then steadies himself by leaning on a staff with his left hand, and feels for the *Solen* [a genus of razor shell] with his naked right foot. This he catches, and holds between his big toe and the next; but although his toes are protected by linen bands, the struggles of the *Solen* to escape are so violent, and the edges of the shell so sharp, that very often a severe wound is inflicted by it."

RED-BACKED VOLE

RED-BACKED VOLES, or red-backed mice as they are sometimes known, are rodents that belong to Muridae, the largest mammalian family. Some scientists place red-backed voles in the subfamily Microtinae along with the bank voles, tree voles, lemmings and other species of rodents.

There are six species of red-backed voles in the genus *Clethrionomys*. Three of these, *C. rufocanus*, *C. centralis* and *C. glareolus*, are exclusive to the Old World; *C. glareolus* is known as the bank vole and is discussed elsewhere in this encyclopedia. *C. rutilus* is found both across the boreal (northern) zone of Eurasia and in northern North America, where it is known as the northern red-backed vole. The southern red-backed vole, *C. gapperi*, is found from Alaska south to the Appalachians and New Mexico. The western red-backed vole, *C. californicus*, ranges from Oregon to California.

Color varies even within litters

A red-backed vole is most easily distinguished by the broad, rust-brown panel of fur down its back, from the head to the rump. The flanks are typically dark gray, with only a faint trace of a rusty hue. The hairs on the belly are dark at the base and white at the tip, producing a pale, frosted effect, and the paws are gray. The tail, which is dark above and pale below, is thin, short, and sparsely haired, and the ears are rounded. There is, however, considerable variation in color among localized populations of red-backed voles. Some may appear almost uniformly dark gray or black with an indistinct stripe, while others, even within the same litter, may have a washed-out, pale buff-gray appearance. The head and body combined can measure up to about 4½ inches (11.5 cm) and the tail can add another 2⅜ inches (6 cm) to this. The red-backed vole weighs up to 1½ ounces (42 g).

Red-backed voles usually build their nests under brush piles, beneath the roots of logs and stumps or directly on the forest floor.

Ground cover, which may be in the form of undergrowth or rocks, offers red-backed voles protection from the eyes of predators and from the elements.

SOUTHERN RED-BACKED VOLE

CLASS	**Mammalia**
ORDER	**Rodentia**
FAMILY	**Muridae**
GENUS AND SPECIES	***Clethrionomys gapperi***

ALTERNATIVE NAMES
Red-backed mouse; Gapper's mouse

LENGTH
Head and body: up to 4½ in. (11.5 cm); tail: up to 2⅜ inches (6 cm)

DISTINCTIVE FEATURES
Rounded ears; small, prominent eyes; short, thin tail; broad, reddish or rufous panel runs from behind eyes to rump; pale gray or white belly and tail underside; gray paws

DIET
All parts of shrubs, grasses and trees; also fungi, roots, invertebrates and carrion

BREEDING
Age at first breeding: 3 months (female), 4 months (male); breeding season: January–November; gestation period: 17–19 days; number of young: usually 5 to 7; breeding interval: 2 to 6 litters per year

LIFE SPAN
Up to 20 months; usually 3–12 months

HABITAT
Mainly cool, moist forests with plenty of ground cover; also rocky terrain, riverside meadows, bogs and tundra

DISTRIBUTION
Eastern Alaska east to Labrador, south into Rockies as far as New Mexico and through eastern states as far as Appalachians

STATUS
Generally abundant; localized scarcity due to habitat loss

Forest rodents

Across their wide range red-backed voles inhabit cool, moist habitats. They prefer mature coniferous or mixed forests with plenty of ground cover in the form of fallen logs, stumps, exposed roots, mossy boulders, brush piles and a carpet of pine needles or leaves. Montane and subalpine spruce, fir, aspen and pine forests are favored, as are mossy rock slides, scree and talus (rock debris at the base of cliffs). The various species also occur in cool, grassy meadows and among riverside stands of willow as well as beyond the tree line on the tundra and on bogland.

A red-backed vole uses runways beneath the vegetation, appropriating those of other small mammals, such as moles or shrews. Where tree squirrels are also present in a particular habitat, the vole uses their middens (dung piles) for shelter and food. Vole populations disappear swiftly from clear-felled or recently burned forest, as such environments give little protection from inclement weather and predators and the clear soil is barren and of little use for foraging.

In summer, the red-backed vole typically roams an area of 3 acres (1.2 ha), although it may sometimes cover an area of up to 7½ acres (3 ha). This range may shrink to one-tenth this size during the winter. The vole remains active through the colder months, scurrying through tunnels beneath the snow. It builds a snug, spherical nest, 3–4 inches (7.5–10 cm) in diameter, out of soft plant parts such as moss and lichen. The nest may be sited in a stump, log, short burrow or tree hole, or directly upon the ground surface. The vole defends the surrounding territory both against others of its kind and against other species, and is aggressive except

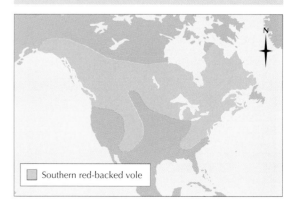

Southern red-backed vole

when breeding. A sharp chirp, audible to humans up to 6 feet (1.8 m) away, may be enough to ward off intruders. The vole is less territorial during winter, when several voles may huddle together in a nest for warmth. At such times the population density can soar to about 75 voles per 2½ acres (1 ha). Extremes of density are also associated with population booms and troughs, which occur randomly, rather than cyclically every few years.

Opportunistic feeders

The red-backed vole adjusts its diet to whatever food is available. In warmer months it eats all parts of shrubs and trees, including tender shoots and the bark, twigs, buds, leaves, flowers, seeds and cones of conifers. It snaps up any insect larvae it turns up in the leaf litter, and even nibbles on carrion. Fruits, berries and fungi of all types, including truffles, enter the diet in the summer and fall. When the cold northern winters constrict the diet, the vole resorts to feeding on tougher plant parts, and continues to forage below the snow for tree roots, bark and seeds. Throughout the year the red-backed vole drinks a large amount of water, up to ¾ fluid ounce (22 cc) per day.

Prolific breeders

Able to breed from mid-January through late November, the red-backed vole is a prolific breeder, with mature females capable of rearing up to six litters per year. A more usual maximum is three or four litters, with peak breeding occurring from February through October. There is no firm pair bond, and a female mates with more or less any male she comes across. After a gestation period of 17–19 days a litter of 2 to 11 young is born in the nest chamber. The average litter size increases with latitude, from around five in the south to seven farther north. Naked and blind at birth, the young are able to stand when they are about 4 days old. They open their eyes no later than 15 days and are weaned off their mother's milk at 3 weeks. Within 3 months female offspring are themselves ready to breed, with males maturing a month or so later.

With each territorial, breeding female able to rear more than two dozen young in a year, populations might explode were it not for the typically brief life span. Many red-backed voles die by the age of 3 months, and few exceed 12 months, despite occasional records of individuals living for nearly 2 years. Cold temperatures and exhaustion claim many lives, but others fall to a host of predators, including weasels, skunks, foxes, cats, coyotes and birds of prey. Fleas, ticks and internal parasites also take their toll.

Red-backed voles are disliked by forestry owners as they can damage tree seedlings. Poison (grain bait laced with zinc phosphide), traps and snares are just a few of the methods used to control red-backed vole numbers, although the species' economic impact is minimal overall. However, red-backed voles can also be a help: they eat harmful insect larvae.

Although red-backed voles are highly territorial and do not live communally or form lasting pairs, mothers and their offspring do form amicable relationships.

RED FOX

A red fox cub in flight. The weeks spent in play with siblings help a cub develop its physique and hunting skills, readying it for a time when it must fend for itself.

IT HAS BEEN SAID THAT but for its careful preservation for hunting purposes, the red fox would have become extinct long ago in the British Isles except in the wildest and most remote corners. For centuries and throughout its range the red fox has been persecuted by humans not only for sport but also because of its poultry-killing habits. Even today the killing of a fox is often approved of by farmers. Yet, in spite of all this, the fox has survived, and at times and in some areas is unusually numerous.

The head and body of the red fox usually measure just over 2 feet (60 cm), with a 1⅓-foot (40-cm) tail. However, there are records that greatly exceed these measurements, especially in Scotland. A well-grown fox stands only about 1⅕ foot (35 cm) at the shoulder. The male fox and vixen (female) are alike except that the vixen is slightly smaller and often has a narrower face. The fur is sandy, russet or reddish brown above and either light cream or dark grayish black on the underparts. The backs of the ears are black, as are the fronts of the legs, but these may be brown and can change from one color to another with the molt. The colors may vary, however, not only between one individual and another but also in the same individual from season to season. The foxes of Scotland, although of the same species, usually have grayer fur than English foxes. When fully haired, the tail is thick and bushy and is known as a brush. The tip (or tag) is usually white but may be black. Red foxes vary considerably in weight, from around 9 to just over 30 pounds (4–14 kg).

The sharp-pointed muzzle, the erect ears and the quick movements of the eye with its elliptical pupil combine to give the red fox an alert, cunning appearance, and many folklore tales of its astuteness have been invented in the past. At the molt, which takes place in July and August, foxes lose their characteristic appearance and look more thin-bodied, long-legged and slender of tail.

RED FOX

CLASS	**Mammalia**
ORDER	**Canidae**
FAMILY	**Vulpes**
GENUS AND SPECIES	***Vulpes vulpes***

ALTERNATIVE NAME
Rufus (archaic)

WEIGHT
9–31 lb. (4–14 kg)

LENGTH
Head and body: 1½–3 ft. (45–90 cm);
tail: 1–1⅘ ft. (30–55 cm)

DISTINCTIVE FEATURES
Sharp-pointed muzzle; erect ears; coat usually sandy, russet or reddish brown; underside either whitish, gray or black; thick, bushy tail (brush), often with white tip

DIET
Small mammals, birds, insects, berries and carrion; garbage and scraps in urban areas

BREEDING
Age at first breeding: 9 months; breeding season: usually late December to early February; number of young: usually 4 or 5; gestation period: about 53 days; breeding interval: 1 year

LIFE SPAN
Up to 12 years

HABITAT
Forest, farmland, hills, tundra, steppe, scrub, bush, semidesert and urban areas

DISTRIBUTION
Throughout Europe and Asia as far south as central India; northern Africa; North America; introduced to Australia

STATUS
Common; subspecies *V. v. necator* of California is very rare and declining

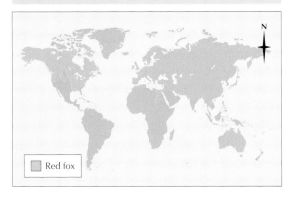
Red fox

The red fox has several color phases. The cross fox, for example, a yellowish brown variety of the species, has a cross-shaped, black band across the shoulders. Another color phase is the silver fox, a variety that has a lustrous black coat with white tips to the guard hairs.

One of the widest distributions

The red fox has one of the greatest natural global distributions of any terrestrial mammal besides human beings. It ranges over Europe and Asia, as far south as central India, as well as into northwestern Africa and Sudan. In Central Asia it lives up to 14,000 feet (4,270 m) above sea level. It is also found across much of Canada and the United States, as well as in Australia.

Adapted to most habitats

The red fox's traditional cunning is a reflection of its adaptability. It favors wooded or bushy areas, but is found in a variety of habitats including deep forest, tundra, farmland, steppe and scrubland. Many foxes today are even found living in or near large towns or cities, where they feed on rats, mice, birds and fruits, and scavenge as well. The red fox has maintained, or even increased, its range across the world in the last 100 years because of its ability to exploit artificial environments. However, the subspecies *Vulpes vulpes necator*, found in the Sierra Nevada of California, is very rare and evidently in decline.

Although the fox lives mainly on the ground, there are many instances of it climbing trees. This usually occurs when a tree is leaning or when

A red fox in winter, Manitoba, Canada. The red fox is one of the most widely distributed of all mammals, partly because it adapts so readily to human environments.

Red foxes hunt at night, preying mainly upon rabbits and rodents. It has been said that the role they play in keeping these animals in check more than offsets the occasional chicken that is taken.

foxes repeatedly visiting poultry farms or private gardens containing poultry and never molesting them.

More solid information about the red fox's diet comes from a British governmental investigation of the stomach contents of dead foxes. This showed that foxes eat mainly rabbits, rats, mice and bank voles. Hedgehogs, squirrels, voles, frogs, even snails and beetles are also eaten, as well as a great deal of vegetable matter. Partridges, pheasants and other birds are also taken. A fox will soon discover offal or carrion, even if buried up to 2 feet (60 cm) in the earth. Foxes also visit gardens and compost heaps, and many of the foxes now living in towns have turned to scavenging garbage and scraps.

Teaching the cubs

Mating varies according to latitude, but takes place from late December to early February in most places. The gestation period is 53 days. The vixen produces her single litter for the year in the spring. There are usually four or five cubs, although there may be up to 13. The young are blind until 10 days old and remain in the earth until nearly 1 month old, the vixen staying close beside them. During this time the male fox often plays a part in supplying the food. When about 1 month old, the cubs come out in the evening and can be seen playing as a group with the parents outside the earth. This continues for several weeks.

there is a trailing bough that has broken and is hanging down to the ground, up which the fox can clamber. Foxes are largely nocturnal, but they can often be seen during the day. Except during the breeding season, the male fox and vixen lead solitary lives. Most of the day is spent in an earth, which is more of a cavity in the ground than a burrow. They may make this themselves or use a badger's set or rabbit burrow.

Foxes use a great variety of calls, the most familiar being the barking of both the male fox and the vixen in winter and the screaming of the vixen, generally during the breeding season. It has now been established that, contrary to common opinion, the male fox also sometimes screams.

Poultry killer?

A great deal has been written about the fox prowling around farms looking for an opportunity to kill an unguarded fowl. Certainly foxes will take poultry and may also take lambs, but these habits tend to be local. A vixen that has taken to killing poultry may teach her cubs to do the same. However, not all foxes are habitual poultry stealers, and there have been instances of

After the cubs are weaned, the male fox continues to bring food for them. The cubs spend their time playing together, and in this way they get exercise, develop their limbs and learn to coordinate movements and senses. Later the vixen takes them hunting at night, so they learn from her example how to fend for themselves. The cubs usually leave their parents when about 2 months old, though some stay longer with the family. They reach adult size 6 months after birth, and become sexually mature in their first winter. Red foxes have been known to live up to 12 years. Paradoxically, hunting may help fox populations remain dynamic, by culling older individuals and enabling younger ones to move more rapidly into vacated territories.

RED PANDA

THE RED PANDA IS the first panda that became known to science, although the name is today more familiarly applied to the larger giant panda, *Ailuropoda melanoleuca*. The red panda was first made known to western scientists in 1825, nearly 50 years before the giant panda became known to Europeans. It is catlike in shape, with a head and body of up to 24 inches (60 cm) in length and a brushlike tail of up to 18 inches (45 cm). It is one of the most richly colored of mammals. The body is covered with chestnut-colored woolly fur, its face is white with a dark stripe from the eye to the corner of the mouth and the tail is ringed. The underside and the limbs are black and the toes bear long, partly retractile claws. The red panda is classed in the Procyonidae, the raccoon family, along with the raccoon, coatis, kinkajou and ringtails (all of which are discussed elsewhere in this encyclopedia).

Mountain forest dweller

The red panda spends most of its time in trees in the high mountain forests of Nepal, northern India, northern Myanmar (Burma) and parts of southern China, at altitudes of more than 16,500 feet (5,000 m). It spends much of the day curled up asleep on its side, in the manner of a domestic cat, on a branch or in a tree hollow. When it comes down to the ground it walks slowly and awkwardly, with the toes slightly turned in and the hind feet plantigrade (with the whole sole of the foot on the ground, as in bears and humans).

The red panda eats primarily leaves and fruits, some of which it finds on the ground. It forages mainly in the early morning and in the evening, although in some areas of its range it may feed all day. It is thought to eat eggs and is reputed to enter local villages at times to help itself to milk and butter. The red panda also takes insects, small birds and small mammals.

Solitary lifestyle

Although red pandas sometimes live in family groups, most of the time they are solitary. They communicate with each other by leaving odor from an anal scent gland on the branches and trunks of trees within their individual territories. Odor is also emitted from this gland if the animal

is excited or agitated. When red pandas are within hearing range of each other, they often make whistles and chirps.

Even though the species has been well studied in captivity, little is known of the ecology of the red panda in the wild. Scientists know that two or three young are born in spring following a gestation of 114–145 days and that these remain with the parents until shortly before the next litter is born. It appears that in some areas red pandas are nocturnal, while in others their lifestyle is more diurnal (day-active). It seems likely that the animals develop more nocturnal habits in those areas where disturbance from local humans populations is greater.

Red pandas have few natural predators. Their enemies include larger carnivores such as leopards and clouded leopards, as well as birds of prey. When red pandas are on the ground they are most vulnerable. The offspring of red pandas

Its mask and ringed tail give the red panda a superficial resemblance to the raccoon. Both species belong to the family Procyonidae.

The red panda spends much of its life in trees. Although it is mainly vegetarian, it also eats insects, bird eggs and small birds.

are particularly vulnerable in this respect, because they are naturally inquisitive and curious about their environment. Female red pandas keep a close eye on their young as they explore their surroundings. If they stray too far from their mother, they are called back by sharp whistling chirps and yelps. Climbing is an important activity, and one that young red pandas learn from an early age, although many die or suffer broken limbs from falls before they have mastered this skill.

An endangered species

In some areas red pandas are locally common. However, globally their numbers are decreasing due to habitat loss and to the fact that they are hunted for their fur. In parts of southern China, red panda fur is used to make hats that form an important part of the traditional marriage ceremony. Moreover, red panda populations recover very slowly from any losses. Females are only able to conceive for one day in the year, and often do not breed every year.

Red pandas require high quality fir and birch forests with a plentiful supply of fruits and bamboo for nourishment. Mature trees with potential den sites in the form of tree holes are also an important requirement for them.

RED PANDA

CLASS	**Mammalia**
ORDER	**Carnivora**
FAMILY	**Procyonidae**
GENUS AND SPECIES	***Ailurus fulgens***

ALTERNATIVE NAMES
Lesser panda; cat-bear; fox-cat, fire-fox

WEIGHT
6⅔–13¼ lb. (3–6 kg)

LENGTH
Head and body: 20–24 in. (50–60 cm); tail: 11–18 in. (28–45 cm)

DISTINCTIVE FEATURES
Rounded head; large, erect ears; short muzzle; white face and chin; red vertical cheek stripes; chestnut-red shaggy fur on upperparts, longer in colder climates; ringed, red and black-brown tail; dark brown or black legs and underparts

DIET
Mainly roots, grasses, leaves, fruits and tree sap; also bird eggs, small birds and insects

BREEDING
Age at first breeding: 18 months; breeding season: January–March; number of young: 2 or 3; gestation period: 114–145 days; breeding interval: 1 or 2 years

LIFE SPAN
Up to 17 years in captivity

HABITAT
Temperate forest above 16,500 ft. (5,000 m)

DISTRIBUTION
Nepal, northern India, northern Myanmar (Burma) and southern China

STATUS
Endangered; estimated population: less than 10,000

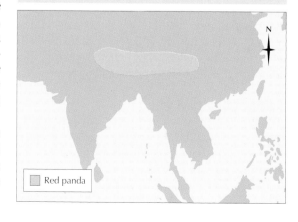

Red panda

REDPOLL

THERE ARE FOUR SPECIES of redpolls, the lesser redpoll (*Carduelis cabaret*), the mealy redpoll (*C. flammea*), the greenland redpoll (*C. rostrata*) and the Arctic or hoary redpoll (*C. hornemanni*). They are all closely related to, and share the same genus as, the twites, siskins and linnets. The lesser redpoll, about 4½–5¾ inches (11.5–14.5 cm) long, is often overlooked by people, being mistaken for a sparrow. A second look shows that it has the same forked tail as the linnet, *C. cannabina*, and close examination shows a crimson forehead and black chin. The upperparts are streaked with two buff wing bars, and in summer the male has a crimson tinge to his breast and rump.

The mealy redpoll was formerly divided by ornithologists into three subspecies, of which the British and montane European subspecies was known as the lesser redpoll. Now ornithologists recognize four true species: the three forms once regarded as subspecies of the mealy redpoll, together with the Arctic redpoll. The mealy and Greenland redpolls are paler than the lesser redpoll. The Arctic redpoll has a white rump and underparts and much paler upperparts.

Mealy redpolls breed in northern North America, northern Asia, Scandinavia and Iceland. Lesser redpolls breed in the British Isles and the Alps, and were introduced into New Zealand in the latter half of the 19th century. The Greenland redpoll breeds in Greenland and Iceland and has occasionally bred in Scotland. The Arctic redpoll breeds in northernmost North America, Asia and Europe.

Flocking together

Outside the breeding season redpolls live in small flocks, flying from tree to tree with the bouncing flight of finches and keeping in touch with a twittering flight call usually rendered as *chuch-uch-uch-uch*. Flocks of redpolls often combine with flocks of siskins and other small birds, and the calls of these birds seem to be similar enough to be mutually understood. For example, they use similar alarm calls to signal the approach of an aerial predator such as a sparrow hawk, *Accipiter nisus*.

Redpolls live on small seeds and small insects, feeding among trees or on the ground. The main food of the lesser redpoll is the seeds of birch and related trees. Seeds of grasses, willow herbs, thistles, willows, conifers and several other plants are eaten when birch seeds are not available. In the spring, when there are few seeds, small animals are taken, especially when nestlings are being fed. This animal food consists mainly of insect larvae, caterpillars and aphids taken from leaves or buds. Buds of birch and other plants are sometimes eaten when there is a shortage of other food. Sometimes redpolls attack the blossoms of fruit trees in such numbers that they severely damage the crop.

Butterfly-like display

The redpolls' song is a feeble trill interspersed with call notes, and the males often display in the air, circling and looping with slow wing beats, like butterflies. They also display in trees, hopping from branch to branch with tails spread and wings raised over the back. The nest has a rough foundation of twigs on which is built a cup of grasses and moss, neatly lined with down and feathers. Nests may be found in hedges, trees, bushes or osiers, and occasionally in coarse grass or heather. The Arctic redpoll nests in birch

A male mealy redpoll photographed in the Swedish winter, its plumage fluffed up against the cold.

Redpolls move south in late autumn when their food supplies run short. They travel farther in years when the seed crop is poor.

or willow herb and even on the ground. The female builds the nest and incubates the eggs and is fed by the male while she is sitting. The usual clutch is four to six pale blue eggs with light brown markings. The chicks hatch after 10–12 days, are fed by both parents and fly after 9–14 days. Two clutches per year is the norm.

Migration

Most redpolls, particularly Arctic redpolls, migrate southward in the autumn. Mealy redpolls sometimes reach the Mediterranean region. Recoveries of banded birds show that redpolls that have nested in northern Britain do not migrate straight across the North Sea to mainland Europe. They fly south and then cross the Channel at its narrowest point, from southeastern England.

The numbers of redpolls that migrate seems to depend very much on food supply, as does the distance traveled. Even the extra feeding time given by the extra minutes of daylight a few hundred miles farther south may make all the difference to survival in winter when food is scarce. Some redpolls spend the winter in England, especially if the birch crop is good. However, it has been shown that more redpolls banded in Britain are recovered in mainland Europe in years when there is a poor birch crop. Migration and food supply have also been shown to be linked in Scandinavian redpolls.

MEALY REDPOLL

CLASS	**Aves**
ORDER	**Passeriformes**
FAMILY	**Fringillidae**
SUBFAMILY	**Carduelinae**
GENUS AND SPECIES	***Carduelis flammea***

WEIGHT
½–⅔ oz. (12–16 g)

LENGTH
Head to tail: 4½–5¾ in. (11.5–14.5 cm); wingspan: 8¼–9¾ in. (21–25 cm)

DISTINCTIVE FEATURES
Compact body; conical, sharply pointed bill; small crimson patch on forehead; small black chin; brown, heavily streaked upperparts; pale whitish underparts, with crimson wash in summer; forked tail

DIET
Seeds, especially of trees such as birch; some invertebrates (early summer only)

BREEDING
Age at first breeding: 1 year; breeding season: eggs laid May–July; number of eggs: usually 4 to 6; incubation period: 10–12 days; fledging period: 9–14 days; breeding interval: usually 2 broods per year

LIFE SPAN
Up to 8 years

HABITAT
Forests and woods, particularly with birch, willow and alder trees

DISTRIBUTION
Breeding: across Alaska, northern Canada, Greenland, northern Europe and Siberia; also in mountains of central Europe. Winter: populations move south.

STATUS
Common

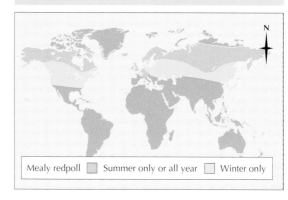

Mealy redpoll ▢ Summer only or all year ▢ Winter only

REDSTART

THE EURASIAN REDSTARTS are small relatives of Old World thrushes and robins. In the breeding season, the male common redstart, *Phoenicurus phoenicurus*, has blue-gray upperparts and a rufous-orange breast, rump and tail. The face and throat are black and the forehead is white. In winter the throat and breast feathers become tipped with white. The female's plumage is dull by comparison with that of the male, being brownish above and buff underneath. The young are mottled like young European robins, *Erithacus rubecula*. The black redstart, *P. ochruros*, is the same size as the common redstart, about 5½ inches (14 cm) long, but the male is almost wholly black and the female is very dark. Both have the characteristic chestnut rumps and tails.

The common and black redstarts are the European representatives of about a dozen species of redstarts, most of which live in Central Asia. One species, Moussier's redstart, *P. moussieri*, lives in North Africa. Very closely related are the rock thrushes of southern Europe and Asia. Redstarts and rock thrushes share the habit of constantly quivering their tails. The American redstarts are wood warblers, unrelated to the Eurasian species, and are discussed elsewhere under that name.

Birds of house and garden

Redstarts are restless birds, constantly flitting from perch to perch, quivering their tails and bobbing up and down. The common redstart spends most of its time in trees, while the black redstart spends more time on the ground and runs rapidly, like a wagtail. These differences are reflected in the habitats of the two species. The common redstart frequents woods, parks, heaths and gardens, whereas the black redstart is found in rocky country and cliffs or near buildings and waste country. In Germany the common redstart is known as *Gartenrotschwanz*, or garden redstart, and the black redstart is *Hausrotschwanz*, or house redstart. Several species of redstarts are found in the Himalayas.

Long-distance migrants

Redstarts are migratory birds. The common redstart travels south across the Mediterranean and the Sahara to spend the winter in the scrub and grassland just north of the equator. The white-capped redstart, *Chaimarrornis leucocephalus*, which nests by mountain streams in the Himalayas up to 16,000 feet (4,800 m) above sea level, winters on the Indian plains.

Mainly insects

Redstarts feed on flying insects, such as small beetles, butterflies and flies, and on ground-living animals, such as caterpillars, spiders, small worms and wood lice. Berries are sometimes eaten, especially in the fall. The nestlings are fed mainly on caterpillars if they are available, including hairy caterpillars, which are ignored by most other birds.

Male redstarts arrive back from the winter quarters shortly before the females and start singing a warbling tune to demonstrate their ownership of territories. They continue to sing until the eggs hatch, except for a period of silence while the hens are building their nests.

The nest is built in a hole or crevice in a tree, stump or wall. Abandoned swallow nests are also used, and redstarts are often attracted to birdhouses. Although the nest is built solely by

Holes in tree stumps are favorite nesting sites for redstarts. The male common redstart (above) uses song and its bright colors to attract the female to a suitable location.

A female common redstart pauses to drink at a pool during its northbound spring migration from Africa.

COMMON REDSTART

CLASS	**Aves**
ORDER	**Passeriformes**
FAMILY	**Turdidae**
GENUS AND SPECIES	***Phoenicurus phoenicurus***

WEIGHT
⁴⁄₁₀–⁷⁄₁₀ oz. (11–19 g)

LENGTH
Head to tail: about 5½ in. (14 cm)

DISTINCTIVE FEATURES
Breeding male: rufous-orange breast, rump and tail; black face and throat; small white mark on forehead; blue-gray upperparts. Nonbreeding male: white tips to throat and breast feathers. Female and young male: rufous rump and tail; brownish upperparts; pale buffish underparts.

DIET
Mainly spiders and insects; also small fruits in late summer and fall

BREEDING
Age at first breeding: 1 year; breeding season: eggs laid late April–May (Europe); number of eggs: 5 to 7; incubation period: 12–14 days; fledging period: 14–15 days; breeding interval: 1 or 2 broods per year

LIFE SPAN
Up to 9 years

HABITAT
Open deciduous or coniferous woodland; also heaths, parks and large gardens

DISTRIBUTION
Breeding: much of Europe, east through Ural Mountains to Mongolia, south to Iran and Pakistan. Winter: parts of sub-Saharan Africa and Middle East.

STATUS
Common

the female, the male occasionally contributes by collecting a few pieces of material. A foundation of sticks and roots is made at the bottom of the hole or crevice, and a cup of grasses, moss, hair, feathers and other materials is built on top.

The eggs, pale blue and sometimes speckled with brown, are laid shortly after the nest is finished. The usual clutch is five to seven in Europe. As with many small birds, the redstarts lay more eggs in the northern parts of their range than in the south. Redstarts have two broods over most of their range, but only one in the south. Incubation lasts 12–14 days and is carried out by the female alone. The male stands guard nearby, singing and attacking intruders such as the wryneck, *Jynx torquilla*, a relative of woodpeckers that also nests in holes in trees.

The chicks hatch out naked and are brooded for several days by the female while the male feeds them. Later both parents bring food to the growing chicks, which leave the nest when they are about 2 weeks old. Their parents continue to feed them for a few days, and the brood keeps together for a few weeks.

Welcoming male

The male redstart's bright colors are displayed to the female during courtship, but they are also used to lure her to a hole that the male has selected as a good nesting place before he has even found a mate. He shows off the white patch on his head and his colorful orange-chestnut rump and tail, drawing attention to the chosen nest hole by climbing in and out. He also flutters about near the hole, singing. If for any reason a nest is destroyed, the male selects a new site and lures the female to it.

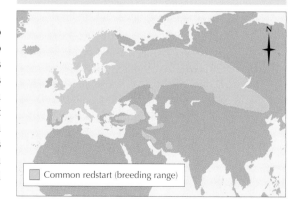

Common redstart (breeding range)

RED-WINGED BLACKBIRD

WHEN EUROPEAN settlers first arrived in North America they named many of the birds they saw after similar-looking Old World species, even if the likeness was only superficial. Consequently, a lively bird with a red breast was christened the American robin, *Turdus migratorius*, and various birds with a primarily black plumage were dubbed blackbirds. In fact, the American robin is more closely related to the European blackbird, *T. merula*, than to the European robin, *Erithacus rubecula*, while the red-winged blackbird is a member of an entirely different family, the Icteridae.

There are 104 species in the Icteridae, and together they make up one of the most important New World bird families. Ranging from Alaska to the southern tip of South America, they include some of the most abundant songbirds in the Americas. Some have become unpopular with humans, for a variety of reasons. The brown-headed cowbird, *Molothrus ater*, is a brood parasite, laying its eggs in the nests of other birds, and it has been blamed for the decline in population of many species of songbirds. Moreover, large congregations of red-winged blackbirds and their allies are responsible for severe crop damage. Not only is the red-winged blackbird the most abundant icterid in North America, it is almost certainly the most abundant bird of any description on the continent. However, not all icterids are common, and some are regarded as vulnerable.

Unusual song

One of the first indications of the presence of red-winged blackbirds is the distinctive ringing, scratchy trill of the males' song during the breeding season. At closer range it becomes clear that the song is actually made up of three syllables: *kon-ka-reeeeee*. It is repeated at regular intervals, and in a suitable habitat several territorial males may sing at the same time. It is usually not difficult to catch sight of the singer, which generally perches on a prominent stem or overhead wire. Males in breeding plumage are rather stocky, medium-sized birds, with an all-black plumage apart from two red shoulder patches, or epaulets. On closer examination a small yellow patch may be seen under the red. Females are not so easy to identify. Their upperparts are brownish and heavily streaked with black, while the underparts are black-streaked buff. The bill is that of a typical icterid: thick-based but ending in a sharp point.

Red-winged blackbirds are omnivorous ground feeders. They eat mostly plant food, but also take practically any animal small enough to swallow, including earthworms, insects, snails, frogs, lizards, bird eggs, nestlings and fish.

Breeding cycle

In most of North America the adult male red-winged blackbirds return to the breeding grounds first, typically in late March or April. They then establish territories that they defend against other males. The average size of a territory is about 21,520 square feet (2,000 sq m), although the exact size depends on the habitat. Upland territories are larger than those situated

The red-winged blackbird (male, above) is aggressive and abundant. It is probably the most common bird in North America.

The female red-winged blackbird's streaked plumage helps her to camouflage herself on the nest, which is usually located in thick vegetation.

RED-WINGED BLACKBIRD

CLASS	**Aves**
ORDER	**Passeriformes**
FAMILY	**Icteridae**
GENUS AND SPECIES	*Agelaius phoeniceus*

LENGTH
Head to tail: about 8¾ in. (22 cm)

DISTINCTIVE FEATURES
Stocky, medium-sized songbird; rounded wings; shortish tail; moderately thick bill. Male: black, except for orange-red shoulder patches. Female: brown upperparts with black streaks; buff underparts, heavily streaked with dark brown.

DIET
Wide variety of plant and animal matter, including seeds, insects, worms, snails, small reptiles and amphibians and bird eggs

BREEDING:
Age of first breeding: 1 year (female), 2 years (male); breeding season: mainly April–early July; number of eggs: 3 or 4; incubation period: about 13 days; fledging period: 12 days; breeding interval: 1 year

LIFE SPAN
Up to 16 years; usually up to 3 years (female) or up to 4 years (male)

HABITAT
Breeding: marshes, wetlands, damp and shrubby fields, woodland margins and urban parks. Winter: mainly agricultural land, particularly open fields.

DISTRIBUTION
Breeding: throughout North America, apart from northern Canada and Alaska, and south through Central America to Costa Rica. Winter: northern populations move southward.

STATUS
Common to abundant

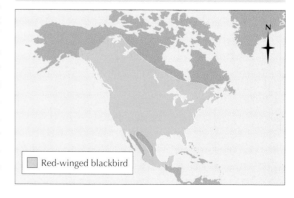

☐ Red-winged blackbird

in marshes. Typical breeding habitat includes cattail, sedge and salt marshes, wet shrubby fields, the margins of second-growth woodland and even urban parks. The use of upland habitats by red-winged blackbirds seems to be a relatively recent development.

Females generally breed from the age of 1 year, but males start a year later. Although physiologically capable of breeding at this age, they only rarely establish territories, and even less frequently have any breeding success. Most males do not even acquire their full breeding plumage, including epaulets, until they are 2 years old.

Red-winged blackbirds are polygamous (taking more than one partner). Males often mate with several females, up to a maximum of about 15. Females also sometimes mate with more than one male. This appears to be not because males are invading their neighbors' territories, but because females sometimes go in search of other males. Females seem to choose a partner based on the qualities of his territory, rather than his personal attributes. The first female to settle in a territory has a dominant position over later arrivals. Although there is still debate about this point, there is some evidence that females establish subterritories within the male's territory.

The male's prime display is called the song spread, in which he fluffs up his body feathers, raises his epaulets and spreads his tail as he sings. The most intensive song spreads are reserved for territory defense against other males. A territorial male red-winged blackbird also performs a flight display, flying over his territory with a fluttering motion, spreading his tail, raising the epaulets and singing as he does so. The clutch size averages slightly more than three eggs north of Mexico, a little less in Central America. The female incubates the eggs alone for up to 13 days and the chicks take about 12 days to fledge. She takes on most of the responsibility for feeding the young, but the male helps a little, particularly with first broods.

The red-winged blackbird is one of up to 250 species of songbirds in North America that fall victim to brood parasitism by brown-headed cowbirds. Females of the latter species visit a red-winged blackbird's nest when the mother is absent, lay an egg, and usually remove one of the mother's eggs. The red-winged blackbird female loses one of her clutch and raises a cowbird, but her own nestlings generally grow as they would have if the nest had not been parasitized.

Strength in numbers

Red-winged blackbirds form communal roosts year-round, but during the breeding season the roosts are relatively small, occupied by first-year and other nonbreeding males. As the breeding season draws to a close other birds join the roosts and begin their fall molt. The red-winged blackbird is a partial migrant, so more northerly populations move south after, or in some cases during, their molt. The southward movement towards milder wintering quarters starts as early as August with some northern Canadian populations and later with more southerly breeders. Female red-winged blackbirds tend to migrate greater distances than males, and consequently male birds predominate in the northern fringe of the winter range.

Red-winged blackbird migration can be highly visible: it is diurnal (takes place by day) and can involve large numbers of birds on the move together. Huge winter roosts, sometimes consisting of millions of red-winged blackbirds, cowbirds, grackles and European starlings, become occupied from November. The red-winged blackbirds may travel up to 48 miles (80 km) from the roost to feeding areas, although 12 miles (20 km) is a more usual distance. They are particularly drawn to grain-growing areas, where farmers regard them as pests and take drastic measures to keep their numbers under control. However, despite such actions, the species' population does not seem to have been affected. Indeed, the red-winged blackbird has become more common in Alaska in the last 20 years and has invaded the Mexican plateau.

Red-winged blackbirds are gregarious birds, often migrating in winter flocks that contain several other bird species as well.

REEDBUCK

Populations of reedbuck, such as this common or southern reedbuck, have been greatly reduced in the past because of excessive hunting.

REEDBUCK ARE AFRICAN antelopes, closely related to waterbuck, kob and lechwe. However, reedbuck are smaller than these other species and have a bare patch below each ear and small, simple, forward-curving horns. The common or southern reedbuck, *Redunca arundinum*, is 47–63 inches (1.2–1.6 m) at the shoulder, with a large bare muzzle. It is light grayish fawn, tawny on the neck and white below. The mountain reedbuck, *R. fulvorufula*, is 43–55 inches (1.1–1.4 m) at the shoulder. Its horns are shorter than its head and are only slightly hooked forward at the tips. The tail is rather bushy. The color varies from grayish fawn to bright red, always with a reddish tinge and white below. The Bohor reedbuck, *R. redunca*, is the same height, but its horns are strongly hooked forward at the tips and are longer than the head. It is a yellowish color, with the neck and head not contrasting with the color of the body, unlike the common reedbuck.

The common reedbuck is found in much of the savanna belt of sub-Saharan Africa. The mountain reedbuck is confined to three isolated populations: one in southeastern Africa; the second in Kenya, Sudan and Ethiopia; and the third centered on the Adamawa highlands of Cameroon. The Bohor reedbuck is also found in savanna, from Senegal east to Ethiopia.

Bounding antelopes

Reedbuck are preyed on by lions, leopards, hyenas, wild dogs, jackals, pythons and eagles. When a herd is threatened, the male Bohor reedbuck stands broadside to the intruder, one hind foot in advance of the other, with his head turned, sniffing the air and watching. Then he bounds away with great leaps. The does (females) retreat first, scattering widely, with the buck (male) following on.

When running, reedbuck hold their tails tightly down between their legs. They bound with the forefeet out in front and the hind feet straight out behind. Mountain reedbuck also scatter when they are disturbed, but common reedbuck run off in a group with their tails held upright, exposing the white undersurface. After they have run 300–400 yards (270–360 m), the groups of reedbuck stop and turn to check if the danger has passed.

REEDBUCK

CLASS **Mammalia**

ORDER **Artiodactyla**

FAMILY **Bovidae**

GENUS AND SPECIES **Common reedbuck,** *Redunca arundinum*; **Bohor reedbuck,** *R. redunca*; **mountain reedbuck,** *R. fulvorufula*

ALTERNATIVE NAME
Southern reedbuck (*R. arundinum*)

LENGTH
Head and body: 43–63 in. (1.1–1.6 m); shoulder height: 24–41 in. (60–105 cm); tail: 6–18 in. (15–45 cm)

DISTINCTIVE FEATURES
Medium to large size; usually tawny, fawn or yellowish with paler underparts; horns (male only) curve forward at tip; black patch on cheeks (*R. arundinum* and *R. redunca* only)

DIET
Grasses, shoots and leaves

BREEDING
Age at first breeding: 9–24 months; breeding season: all year, with peaks at certain times; number of young: usually 1; gestation period: 240 days; breeding interval: 9–14 months

LIFE SPAN
Up to 18 years in captivity

HABITAT
Grassland, open woodland, reedbeds and mountain plateaus

DISTRIBUTION
***R. arundinum*: sub-Saharan savanna belt. *R. redunca*: Senegal east to Ethiopia. *R. fulvorufula*: 3 isolated populations.**

STATUS
Dependent on conservation programs

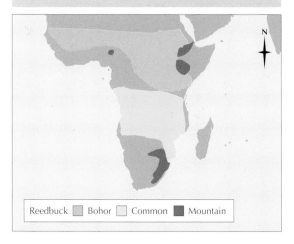

Reedbuck ▨ Bohor □ Common ■ Mountain

Small herds

Bohor reedbuck are found in small groups, usually one male with three or four females. Common reedbuck associate in groups of six or so, and mountain reedbuck are often in groups as large as 20. The mountain reedbuck lives among bushes and dry grass on hilly slopes, coming down at night for water and returning to the higher slopes in the morning. All three are both grazers and browsers. The young, known as lambs, are usually born singly, although very rarely the mother gives birth to twins; they are born year-round. Lambs are born with woolly coats. Their only protection is to lie still, and until they are quite well grown the mother leaves them hidden in the undergrowth while she feeds.

Dancing antelopes

A naturalist working in the dense, grassy Rukwa Valley was the first to describe the "dance of the reedbuck." One evening, he noticed some reedbuck in the distance jumping up and down. On approaching the animals, he saw there were about 250 of them so engrossed in their activity that they failed to notice him. About 90 old bucks were standing or lying around the edge of the "dance floor" while the yearlings were dancing. Adult females were among them but not taking part. The young bucks were following the does as they moved about and bounded high into the air, with their legs in a straight line and their bodies and tails held erect. No mating was seen to take place. More and more were bounding eagerly to join in, while others, exhausted, left the arena and lay panting after their exercise.

Scientists are unsure why reedbuck behave in this way. Possibly it is a courtship ritual on a territorial ground. Yet mating is not always seen, and "dancers" leave the arena of their own accord and rejoin it later.

Two subspecies of the mountain reedbuck (above) are now so scarce that zoologists consider them to be endangered.

REED FROG

THERE ARE OVER 200 species of reed or sedge frogs belonging to three large genera (*Hyperolius*, *Leptopelis* and *Afrixalus*) and 11 smaller ones. Most live in sub-Saharan Africa and Madagascar, but there is one species found only in the Seychelles. Many reed frogs are attractively patterned and colored and can change color in response to temperature or background. They are mostly small species, ranging from about ⅗ to 3¼ inches (1.5–8.2 cm) long, and have expanded toe-pads. The males are distinctive for their very large vocal sacs, which may be two or three times the size of the body when inflated.

Polymorphic frogs

The five-lined reed frog, *Hyperolius quinquevittatus*, from Angola to Tanzania, is a pale, almost golden brown, with five mauve-brown stripes running down its back. These stripes are more distinct in the male than in the female. The painted or marbled reed frog, *H. marmoratus*, often has intricate patterns on its back. However, in common with some other reed frogs, the painted reed frog is polymorphic, meaning that there is extreme variety within the species, and even within populations. For example, there may be striped, speckled or plain individuals. They may also be black and white, black and green, black and yellow, brown and yellow as well as several other variations. This made species identification very difficult until genetic studies were used. The painted reed frog ranges from the South African cape to Zimbabwe and Angola. At the southern end of its range it is green or brown with light green spots, each spot ringed with a narrow black line. By contrast, the rare green reed frog, *H. tuberilingius*, is a plain brilliant green with no markings, but is white on the belly and pink on the hind legs.

The two spiny reed frogs, the small golden spiny reed frog, *Afrixalus spinifrons*, and the brown and white spiny reed frog, *A. fornasinii*, live up to their name, for minute spines can be seen on their heads and backs. They differ from the other reed frogs in that the pupil of the eye is vertical instead of horizontal.

Varied habitats

Habitat is very much dependent on genus and species. Most hyperoliids, for example, are arboreal frogs, usually found in reed margins of lakes and ponds. Those of the *Kassina* genus, on the other hand, live among grasses, but still climb fairly well. Other genera, *Chrysobatrachus* and *Tornierella*, for example, are mostly terrestrial and are often found under rocks, while some

Species identification is sometimes difficult with reed frogs because some species, including the painted reed frog pictured below, exhibit a wide variety of colors and patterns.

REED FROGS

CLASS	**Amphibia**
ORDER	**Anura**
FAMILY	**Hyperoliidae**
GENUS	**14 genera**
SPECIES	**206, including small golden spiny reed frog, *Afrixalus spinifrons*; brown and white spiny reed frog, *A. fornasinii*; arum frog, *Hyperolius horstockii*; and painted reed frog, *H. marmoratus***

ALTERNATIVE NAMES
Sedge frog (all species); marbled reed frog (*H. marmoratus* only)

LENGTH
⅗–3¼ in. (1.5–8.2 cm)

DISTINCTIVE FEATURES
Often brightly colored; expanded toe-pads; some species polymorphic (exhibit extreme variability of appearance within species). Male: very large vocal sacs.

DIET
Insects

BREEDING
Variable; eggs may be laid underwater, on the stems of reeds, or in burrows near water

LIFE SPAN
Not known

HABITAT
***Hyperolius* species: arboreal; found among reed margins of lakes and ponds. *Kassina* species: live among grasses. Other genera mainly terrestrial; often found under rocks. *Leptopelis* species: sometimes semiarid country.**

DISTRIBUTION
Sub-Saharan Africa and Madagascar; *Tachycnemis seychellensis*: Seychelles

STATUS
Many species common

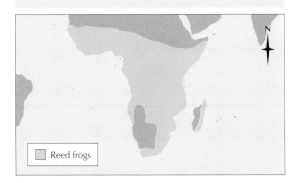
Reed frogs

Leptopelis species live in semiarid conditions and burrow to avoid drying out. Reed frogs are generalist insectivores and feed on flying insects such as mosquitoes. The arum frog, *H. horstockii*, lives in flowers of the arum lily, where its ivory color makes it inconspicuous. It is also overlooked by insects that are attracted by the scent of the arum lily and so fall prey to the frog. When arum lilies are not in flower, the frogs move to other plants and change their color to dark brown to fit their new background.

A pair of painted reed frogs on a floating water plant, South Africa. Species of this genus climb well and are usually found among reeds along the margins of ponds, pools and lakes.

Varied breeding habits

Breeding habitats also vary with species. Reed frogs of the genus *Hyperolius* have small, unpigmented eggs. Those species living in southern Africa spawn among submerged aquatic vegetation, but some West African species attach their eggs to stems of reeds or grasses and the tadpoles slip down into the water after hatching. In the arum frog, clusters of about 30 eggs are laid among water plants, and the surrounding jelly is sticky so the eggs become camouflaged with mud. Once they hatch, the tadpoles feed on minute floating organisms. The painted reed frog lays its eggs in clusters on stones or plants underwater.

In the spiny reed frogs, meanwhile, the eggs are stuck to a leaf, which is folded over to make a protective tube. The small golden spiny reed frog lays its eggs underwater while the the brown and white spiny reed frog lays on plants above the water. Arid-dwelling *Leptopelis* reed frogs spawn in burrows or small depressions near water and the tadpoles wriggle into the water to develop.

REMORA

A live sharksucker rides through the waters off the Bahamas attached to a Caribbean reef shark, Carcharhinus perezi.

THE EIGHT SPECIES OF remoras (family Echeneidae) would be quite ordinary fish but for one thing: where the first dorsal fin should be is a large oval sucker by means of which remoras attach themselves to hosts. The sucker is flat except for its raised edges and ridges across its surface. These are arranged so that the sucker takes a firm hold which can be released only by the remora's swimming forward. As a result, once a remora has fixed itself to a host's body, the forward movement of the host cannot dislodge the passenger, but the remora can voluntarily release itself and swim away. Remoras live mainly in tropical seas. Very occasionally they are found in temperate waters in summer, and then only because they are attached to large hosts that have wandered into the cooler seas.

Ship-holders

The largest remora, the live sharksucker, *Echeneis naucrates*, can reach 3½ feet (1.1 m) in length, whereas the smallest species, the white suckerfish, *Remorina albescens*, is only 12 inches (30 cm) long. Some remoras often attach themselves to sharks, hence the common names for *Echeneis naucrates*, mentioned above, and *E. neucratoides*, the whitefin sharksucker. The common remora,

Remora remora, often, but not always, chooses a shark as a companion too, whereas other remoras prefer different hosts. The whalesucker, *R. australis*, attaches itself only to whales and dolphins, while the marlin sucker (*R. osteochir*) and the spearfish remora (*R. brachyptera*) show a preference for marlins, sailfish and swordfish of the family Istiophoridae. The slender suckerfish, *Phtheirichthys lineatus*, is most often found attached to barracudas.

Remoras also fix themselves to turtles and even the hulls of ships. This last habit led to a belief in ancient times that remoras could stop a ship from moving, which led to them acquiring the name ship-holders. A probable explanation for this belief is that areas of dead water that stop the progress of a ship can occur at sea. When a vessel was becalmed in this way, a man was sent overboard to inspect the hull. The chances of his finding a remora fixed to the hull were high, so the fish was blamed. Some years ago tests were carried out in the New York Aquarium to determine the strength of a remora's sucker. A remora was placed in a bucket of seawater. It promptly fastened itself to the side, and the bucket and its contents, weighing 21 pounds (9.5 kg), could be lifted by holding the fish's tail. A similar experiment with a second fish resulted in a 24-pound (11-kg) bucket being lifted.

Transport only

It was generally accepted that a remora traveling on a shark shared the shark's food but also ate small fish. The current view is that a remora may fasten itself to the shark merely because it is an easy way of traveling, the remora letting go when it sees a shoal of small fish passing by. This may be borne out by the fact that remoras will not share a dolphin's meal. Also, there is little guarantee of food for a remora fastened to a ship's hull. The white suckerfish travels in the mouths or gill cavities of large manta rays (family Mobulidae). They may be helping themselves to their hosts' food, but a more likely suggestion is that they eat these fish's parasites.

Spawning in remoras takes place in June and July in the mid-Atlantic, and in August and September in the Mediterranean. The eggs are 1.5 millimeters in diameter and the newly hatched larvae are 5 millimeters long. The sucker begins

LIVE SHARKSUCKER

CLASS	**Osteichthyes**
ORDER	**Perciformes**
FAMILY	**Echeneidae**
GENUS AND SPECIES	***Echeneis naucrates***

ALTERNATIVE NAME
Striped remora

WEIGHT
Up to 4½ lb. (2 kg)

LENGTH
About 3½ ft. (1.1 m)

DISTINCTIVE FEATURES
Slightly elongated body; flattened head; large sucker situated along top of head and front of body

DIET
Small crustacean parasites infesting skin of hosts; also small free-living crustaceans and fish

BREEDING
Breeding season: early summer; newly hatched fish free-swimming until few inches long, then seeks hosts

LIFE SPAN
Not known

HABITAT
Both close to and far from coasts; often free-swimming in shallow inshore areas; attachments to hosts temporary; hosts include sharks, rays, large bony fish, large sea turtles, whales, dolphins, even ships; may follow divers

DISTRIBUTION
Western Atlantic: Nova Scotia south to Brazil and Uruguay

STATUS
Locally common, especially in warm waters

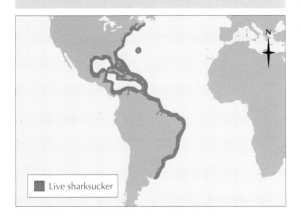

Live sharksucker

to show when the young fish is ¾ inches (19 mm) long, and by the time the remora has grown to 1½ inches (38 mm) long, it can use its sucker disc to hitch a ride.

Turtle-hunting remoras

Remoras are used by some peoples as a fishing aid. A line is tied to the caudal peduncle of the remora, which is then released. The remora seeks out another fish as host, and when it attaches itself both fish are hauled into the boat. From the earliest times, remoras were also used for catching turtles, in places as far apart as Central America, Japan, East Africa and northern Australia. Kenyan turtle fishers try to catch their remoras from large fish, such as snappers (family Lutjanidae), which they catch on a hand line. If this method fails, they fish for remoras directly with baited lines. The remoras are kept in a small stockade in shallow water until needed. When out after turtles, the fishers put them in a basket hung in the water at the stern of the boat.

On arrival at the place where turtles are likely to be, a remora is put into the sea and controlled by a line tied to a ring around its tail. When the remora has fastened onto a turtle, the fisher must play his line skillfully, not jerking it at all. The boat must then be carefully brought over to the turtle. A small metal keeper ring is clipped around the line, and a heavy line with a grapnel at its end is run through this. The grapnel, with four sharp claws, is let down, guided by the keeper ring. When it touches the turtle, it is allowed to drop beyond it and then it is jerked sharply upward. Once one of the claws of the grapnel engages, the turtle is hauled aboard by three or four men, depending on the size of the turtle.

A remora of the genus **Remora** *makes its way through the water without the aid of a host. The common remora,* R. remora, *is the largest of the genus.*

RHEA

Grassland extends down into the far south of Argentina and Chile. Despite the cold here, lesser (Darwin's) rheas find this to be a suitable habitat.

THE FLIGHTLESS RHEA IS THE largest American bird. Its head and neck are almost featherless, its body plumage is soft and its wings are longer than those of the ostrich, which it resembles. The long, powerful legs have three toes each, compared with the two toes of the ostrich. The sexes are similar but the males are a little taller.

There are two species of rheas, of which the greater or common rhea is the more abundant, and is distributed more widely. It is brownish gray above, whitish below and bluish gray on the flanks. There are patches of black on the head and neck. Pure white, albino individuals are quite frequently seen. The greater rhea ranges from northeastern Brazil south to central Argentina. The lesser or Darwin's rhea is smaller and its plumage is brownish with white spots. It ranges from southern Peru to Patagonia in southern Argentina.

Tripping rheas

Rheas live in flocks of 20–30, and rarely up to 100, but old males are always solitary. They live in the open country of the pampas and highland plateaus, where they escape from danger by running, or by crouching if there is enough cover. When running fast, rheas hold their necks out horizontally. They are extremely agile and can change direction with ease, holding one wing out like a sail when cornering.

The flocks sometimes mix with herds of bush deer or guanacos and even with cattle in areas where they are not molested. The rheas are, however, less abundant than they once were. The spread of agriculture has destroyed parts of their habitat and they are often hunted. When horses were introduced by the Europeans, people began hunting rheas from the saddle, bringing the quarry down with bolas: three stones joined by thongs which were thrown at the quarry to entangle its legs. Recently, the decline of the rhea has been hastened by commercial hunting and the colossal export of skins, with over 50,000 traded in 1980. Most of these skins originated from Paraguay, and the main importers were Japan and the United States. Both species of rhea are now in serious decline and are classified as near-threatened.

Rheas are omnivorous, eating both plants and animals. They pluck leaves and seeds or grub roots and catch insects, mollusks, worms and lizards. They also have a habit of snapping up bright objects.

Broody fathers

At the beginning of the breeding season there is intense rivalry between the males. They threaten each other with neck lowered into a U shape. If both hold their ground they fight, twisting their necks together and biting and kicking. When one is vanquished and breaks away it is chased by the victor, which spreads its wings and holds its neck in an S curve.

The victorious males court the females, which live in small groups. The first indication that a male is about to start courting is the ruffling of his neck plumage. The ruffling later extends over his body, and he runs to and fro with his neck stretched and wings spread. At the same time he utters a two-note roar, more like that of a mammal than a bird. He approaches a female with neck lowered and wings spread, drooping to the ground. If the female responds to this display, mating takes place.

Male rheas mate with several females (a practice known as polygyny), and each female mates with a number of males (this is known as polyandry). Rheas are, therefore, both polyandrous and polygynous, but since no lasting pair bond is formed, it might be better to describe them as promiscuous. Nesting and care of the eggs and young is left entirely to the males.

GREATER RHEA

CLASS	**Aves**
ORDER	**Rheiformes**
FAMILY	**Rheidae**
GENUS AND SPECIES	***Rhea americana***

ALTERNATIVE NAME
Common rhea

WEIGHT
44–55 lb. (20–25 kg)

LENGTH
Head and body: 4–5 ft. (1.27–1.5 m)

DISTINCTIVE FEATURES
Very large; long neck; long, thick legs; variable gray-brown overall; feathered thighs; male slightly larger and darker

DIET
Omnivorous: leaves, seeds, roots, fruits, insects and small vertebrates

BREEDING
Age at first breeding: 2–3 years; breeding season: eggs laid August to January; number of eggs: 11 to 18; incubation period: 35–40 days; breeding interval: 1 year

LIFE SPAN
Not known

HABITAT
Open woodland or grassland with tall vegetation

DISTRIBUTION
Northeastern Brazil south to Argentina

STATUS
Near-threatened

☐ Greater rhea

The rhea, this one a lesser rhea in Tierra del Fuego, might look superficially similar to the ostrich, but basic differences between the birds are so great that they are classified in different orders.

The nest is merely a shallow hollow in the ground lined with grass. It is about 3 feet (90 cm) across and 1 foot (30 cm) deep. Each female lays 11 to 18 eggs, depositing them in several nests or even on bare ground if she is ready to lay before the males have built their nests. At first a male leads one female after another to his nest to lay their eggs. Once the male starts to incubate these, he hisses and snaps at any females coming near him. They have to persevere before the male will get up and allow them to lay further eggs.

One male may incubate up to 80 eggs but the clutches usually have 10 to 30 eggs. The chicks soon leave the nest after hatching, in the care of the male. They keep in contact with each other by plaintive whistles, but if one gets lost, it joins another brood. As a result, a family group may consist of young at different ages, and parentage is even more muddled.

What are their relatives?

The large flightless birds such as rheas, ostriches, kiwis, cassowaries, emus and the extinct moas and elephant birds are often grouped together as ratites. At one time these birds were placed in a single family, Ratitae, on the basis of similar characteristics, such as powerful legs for running, small wings and the lack of a keel on the breastbone, which in flying birds is the anchor for the powerful flight muscles. More recently, it has been agreed that although all these birds are descended from ancestors that could fly, far from belonging to a single family, they form four separate, living orders: Struthioniformes (the ostrich), Casuariformes (cassowaries and the emu), Apterygiformes (kiwis) and Rheiformes (rheas). Their similarity is due to convergent evolution because they have all adopted running on the ground as a way of life. The term *ratite* is, however, still used as a loose term for these rather similar birds, as opposed to *carinate*, a term used for any bird that has a keel on its breastbone.

RHINOCEROS

A black rhino, pictured here in Kenya. It has been estimated that 90 percent of adult deaths in this species are due to poaching for rhino horn, which is valued in Chinese medicine as a natural cure for fever and as an aphrodisiac.

TODAY RHINOS ARE A vanishing breed due almost exclusively to relentless killing by humans. These animals are unmistakable in appearance. Like its relatives the horse and the tapir, the rhinoceros bears the weight of each leg on a single central toe, but it has two subsidiary toes on each foot. It differs from tapirs (discussed elsewhere in this encyclopedia) in its bulky build, its thick skin, which is naked or sparsely haired, and the one or two horns on its nose.

There are five species of rhinos in four genera. They actually differ more from one another than, say, horses from zebras. The Sumatran rhino, *Dicerorhinus sumatrensis,* is up to 4¾ feet (1.45 m) high and weighs 750–2,000 pounds (340–900 kg). It has marked skinfolds on the front half of its body only, is distinctly hairy throughout life and has two very short horns. The greater Indian rhinoceros, *Rhinoceros unicornis,* by comparison is up to 6¼ feet (1.9 m) high and some 2½ tons (2,270 kg) in weight. It has a deeply folded skin studded with raised knobs, not much body hair except a stiff brush on the tail tip, and a single horn. The third Asian species, the Javan rhino, *R. sondaicus,* is up to 5¾ feet (1.75 m) high

and weighs perhaps 2 tons (1,800 kg). This species has a less heavily folded skin with differently arranged folds in the shoulder region, and the skin is broken up by a network of cracks into a mosaic of polygons, with a short hair in the center of each, at least in the young. It also has only one horn, which is very short and may be lacking in the female. Both the Javan and Sumatran rhinos used to be found widely throughout Southeast Asia. However, since 1970, more than 84 percent of the world's rhinos have disappeared and only an estimated 12,500 individuals of all species now remain.

The African rhinos are quite hairless except for the tail tip, ear rims and eyelashes, although the young of the white rhino, *Ceratotherium simum,* are quite hairy up to the age of 4 months. They also differ from the Asiatic rhinos in having two long horns, placed closely one behind the other; in lacking front teeth; and in having much less deeply folded skin. The two African species are both plains-dwellers. The black rhino, *Diceros bicornis,* is up to 6 feet (1.8 m) high. Its skin is rough, with grooves over the ribs, and the lip is pointed and prehensile. The white rhino can be just over 6 feet (1.85 m) high, has a smoother skin with no rib grooves, a hump on the shoulder and a long head with a wide, square mouth.

Scent signals

All rhinos are nearsighted but have acute senses of smell and hearing. They tend to deposit their dung in communal heaps and each individual that passes adds to the pile, until heaps 4 feet (1.2 m) high and 20 feet (6 m) across are formed.

Black rhinos tend to be solitary, with 80 percent of adults and 50 percent of weaned calves being seen alone. However, young rhinos often seek the company of one or two other immature animals. The size of an individual's home range varies according to the environment. In the fertile Ngorongoro crater, Tanzania, for example, the average is 6 square miles (15.5 sq km), although immature rhinos have bigger ranges. In the drier habitat of Olduvai Gorge, also in Tanzania, it is 8½–14 square miles (22–36 sq km), whereas in the subdesert climate of Tsavo National Park the rhinos are purely nomadic and no proper home range has been observed.

RHINOS

CLASS **Mammalia**

ORDER **Perissodactyla**

FAMILY **Rhinocerotidae**

GENUS AND SPECIES **White rhino,**
Ceratotherium simum*; black rhino, *Diceros bicornis*; Sumatran rhino, *Dicerorhinus sumatrensis*; Javan rhino, *Rhinoceros sondaicus*; greater Indian rhino, *R. unicornis

WEIGHT
2,200–7,715 lb. (1,000–3,500 kg)

LENGTH
**Head and body: 6½–13¾ ft. (2–4.2 m);
shoulder height: 3¼–6½ ft. (1–2 m);
tail: 2–2½ ft. (60–75 cm)**

DISTINCTIVE FEATURES
Very large body; short, stocky legs; short neck; head adorned with 1 or 2 horns; thick, wrinkled skin; scant coating of hair

DIET
Grasses, leaves and other vegetation

BREEDING
Age at first breeding: 7 years (female), 10 years (male); breeding season: all year, but births may peak at certain times; number of young: 1; gestation period: 420–570 days; breeding interval: often 2 years

LIFE SPAN
Up to 50 years

HABITAT
Savanna, dense forest or shrubby regions. African species: open habitats. Asian species: forests.

DISTRIBUTION
Eastern and southern Africa; tropical Asia

STATUS
Greater Indian rhino: endangered; other species: critically endangered

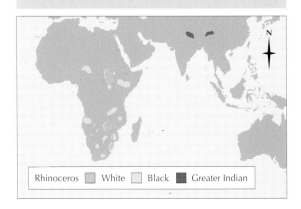

Rhinoceros ▢ White ▢ Black ▮ Greater Indian

Rhinos are active mainly in the morning and evening. They wallow in mud in the evening to cool off. In Ngorongoro their movements are very regular, and they can often be seen at the same time in the same place every day, but in Tsavo the black rhino wanders along well-worn paths over a much wider range and goes to water only every 4–6 days.

Social interaction

The home ranges of rhinos overlap very widely. When two males or two females meet, they are wary of each other but may form temporary associations. A male approaches a female with short, cautious steps, often swinging his head from side to side. The female may approach him or even charge him, in which case he wheels around, gallops off and circles to approach again. This may continue for several hours. When an unfamiliar rhino enters an area where several residents have their home ranges, the residents threaten the newcomer with lowered heads, rolling eyes, flattened ears and raised tails.

A greater Indian rhino, also known as the Asian great one-horned rhinoceros. While the two African species are most often found in open habitats, the Asian rhinos usually live in forests.

Now and again one of them curls its lip and gives vent to a shrill groaning sound while the stranger remains silent. One of the residents will charge it, stopping short in front, and the two will tussle, using their horns as clubs or pikes. If the stranger retreats, it will be pursued, but it may eventually establish itself in the area.

A rhino, on coming up to a dunghill, sniffs it, may push its horn into the dung and then shuffles through it with legs held stiff. After defecation, the dung is scattered with kicks from the hind legs. This smears its feet with its own scent and that of its neighbors.

Black and white rhinos

The black rhino, sometimes called the browse rhino, eats the shoots and twigs of low-growing bushes such as mimosa and buffalo thorn. This is dry, thorny food, which it gathers with its prehensile upper lip. It often feeds at night. It is a nervous animal, easily startled, and what may be seen as aggression could easily be the stampede

Both the Sumatran rhino (pictured below) and the African black rhino have a reputation for being aggressive and unpredictable. This applies particularly to males in the breeding season and to females with young.

of a frightened animal. There is often, however, a real charge when the rhino gallops, head up, toward the intruder, lowering its head at the last moment to strike with its horns. When pressed, the black rhino can run at 35 miles per hour (56 km/h), but it normally charges at 20 miles per hour (32 km/h).

By contrast, male white rhinos are much more gregarious and far less truculent, although the females are less gregarious, often found accompanied only by their calves. However, two or more females with their young may move around together. White rhinos live in low-lying plains. Sometimes called the grass rhino, they feed entirely on grass. They are far less wary than the black species, and so have been quickly exterminated over most of their range. When alarmed, they curl their tails in a loop over the rump and lumber away, reaching a trotting speed of 18 miles per hour (30 km/h). They can also canter at 25 miles per hour (40 km/h).

Rhinos in Asia

The greater Indian rhino lives in the 12-foot (3.6-m) high elephant grass, through which it makes well-worn tunnels. The waterlogged meadows and swamps may be divided into territories or used as common land. A common bathing pool may be used simultaneously by as many as nine rhinos of all ages and both sexes, with the majority lying jammed together in the pool. The dunghills of these rhinos are near wallows or pools, or mark the entrances of tunnels. Unlike the African species, Indian rhinos have glands in the forefeet that leave scent trails in addition to their dung trails. They eat young grass, twigs and water plants, especially water hyacinth, which often forms a carpet on the bathing pools. They run with the tail held down, reaching speeds of up to 25 miles per hour (40 km/h). These rhinos are active during the day but sleep, hidden in the high grass, from midnight until dawn and again at midday.

Javan rhinos, meanwhile, live in deep forest, preferring areas of secondary growth. They feed mainly on young saplings, pushing against them until they break, then eating only a little from the crown. The Javan rhino, especially the male, has a home range, which it leaves every day on wide, regular wanderings. During the rains they bathe and wallow in streams, but in the dry season they use moist, low-lying places or larger rivers, estuaries and

A white rhinoceros cow with her calf, Mkuzi Game Reserve, South Africa. It is thought that captive breeding programs may offer the only hope of survival for some rhino species.

the sea. Dung is deposited in large heaps, as in the Indian rhino, but is often more scattered and sometimes is deposited in streams. Bulls have another method of scent-marking, spattering pungent, orange urine over the bushes.

The habits of the Sumatran rhino seem to be similar to those of the Javan rhino. It eats branches and shoots, leaves, fruits, lichens and fungi. It has a reputation for living at high altitudes and has been found as far up as 8,580 feet (2,615 m) in Java, Indonesia.

Nursed for a year

Breeding generally occurs throughout the year, but births may peak at certain times. Before mating, male rhinos may fight over a female. In the black rhino one will charge, halting suddenly 5 yards (4.6 m) away, and the two joust or try to club each other with their horns. For the white rhino, these fights may be lethal, although it is rare for a rhino to die from its injuries.

The gestation period averages 455 days for the black rhino, 486 for the Indian rhino and 547 for the white rhino. Birth takes about 15 minutes, and the infant is able to stand after about an hour. Only in the white rhino have twins ever been recorded. The calf may begin to graze at one week but suckles for a least a year. Young African rhinos are 1⅔ feet (50 cm) high and weigh 44–55 pounds (20–25 kg) at birth. Young Asian rhinos, meanwhile, stand about 2 feet (60 cm) at the shoulder and weigh up to 50 pounds (23 kg).

A small place for the front horn is visible at birth, the horn beginning to grow in about 5 weeks and reaching 1½ inches (4 cm) long in 5 months. Females are sexually mature at 7 years, males at 10. Rhinos have been known to live up to 50 years.

Hunted for their horns and blood

In Africa, lions take young rhinos when they can, even occasional adults, and adults caught in snares or stuck in the mud have been mauled by hyenas. There is also supposed to be a traditional enmity between rhinos and elephants. When threatened, African rhinos attack with their long horns while Asian rhinos fight more with their tusks. The Sumatran rhino uses its teeth but is said occasionally to toss and trample its victims. Both the black rhino and the Sumatran rhino have received a not altogether fair reputation for being somewhat aggressive and ill-tempered.

However, it is hunting pressure by humans that poses the greatest threat to rhinos, their blood and horns being taken for use in Chinese medicine. The horn is taken by humans to reduce fever or as an aphrodisiac. Among black rhinos, 90 percent of adult deaths are due to poaching. As a result, the Sumatran rhino is now considered critically endangered, as are the Javan, white and black rhinos. The greater Indian rhino is classed as endangered. In some areas rhinos are now being moved into specially fenced sanctuaries for protection from poachers, and many countries now ban the trade in rhino horn.

RICE RAT

THERE ARE ABOUT 81 species of rice rats, all belonging to the same family of rodents as the hamsters and voles. Rice rats are mouselike animals with naked ears. The head and body length ranges from 2½–8 inches (62–200 mm) and the tail from 3¾–4 inches (94–104 mm). The coarse fur is grayish brown to tawny, mixed with black on the back, paler on the flanks and whitish on the underparts. Some species have long tufts of bristles on the hind feet, which project beyond the claws.

The many species of rice rats range over North America from New Jersey and Illinois in the United States through Central America and most of South America. Of the four species that were originally native to the Galapagos Islands, three are now extinct. The remaining species, itself endangered, represents the only indigenous land mammal on these islands apart from the Galapagos bat.

Closely related to rice rats are the now-extinct giant rice rats, genus *Megalomys*, of the Caribbean. The three species, about 14 inches (35 cm) long, excluding the tail, lived on the islands of Saint Lucia, Barbuda and Martinique. They were eaten by indigenous Caribbean peoples and hunted by European settlers because they attacked crops. Changes in habitat as well as the introduction of predators such as cats, mongooses and aggressive rat species contributed to their extinction in the early 20th century.

Defeated by immigrants

Rice rats are fairly gregarious animals and are active at intervals throughout the day and night. They occupy a variety of habitats, each species preferring one particular habitat, so that several species can occupy an area without competing with each other. In Belize, for example, the species *Oryzomys couesi* lives outside the forests, except where these are penetrated by grassy tracks, in grasslands, marshes and bush country. *O. melanotis* lives in dense forests and *O. alfaroi* is found in woodlands. Although they do not make runways through low vegetation, rice rats can be tracked down by their feeding platforms and by their nests of woven grass, constructed in clumps of vegetation or in burrows.

Rice rats are agricultural pests and carry diseases such as leishmaniasis, but their habits are inoffensive and they do not attack as common rats do. Their mild nature has probably been the cause of their downfall, because they have retreated from many places where other rodents have been introduced. In the Galapagos Islands, rice rats became extinct where house mice and ship rats were introduced. In parts of South America rice rats used to live in houses but have been driven out by introduced rats. In Jamaica the cane-piece rat, *O. antillarum*, was once regarded by farmers as a pest of sugarcane, but it may now be extinct due to predation by introduced mongooses.

Occasional pests

Rice rats eat mainly grasses and sedges, cutting them down and eating just the succulent parts. They also eat fruit, seeds and some small animals. Rice rats are often considered pests like other rodents, since they invade crops of sugarcane, rice and maize. They knock down sugarcane and eat it where it lies, whereas newly sown maize is found by burrowing, hence the Brazilian name of *rao mineiro*, or mining rat. Rice rats normally do little damage to crops and only become pests when their numbers build up to plague proportions.

Nesoryzomys bauri is the only rice rat found on the island of Santa Fé, part of the Galapagos Islands. It nests in the hollow trunks of the Opuntia cactus and beneath rocks, coming out to feed at dusk.

RICE RATS

CLASS **Mammalia**

ORDER **Rodentia**

FAMILY **Muridae**

SUBFAMILY **Murinae**

GENUS AND SPECIES **81 species, including rice rats, *Oryzomys*, 37 species; and Galapagos rice rat, *Nesoryzomys*, 1 species**

WEIGHT
Varies widely according to species: ½–2¼ oz. (15–65 g)

LENGTH
Varies widely according to species. Head and body: 2½–8 in. (62–200 mm); tail: 3¾–4 in. (94–104 mm).

DISTINCTIVE FEATURES
Brown, gray or black upperparts; paler or sparsely furred underparts; tail unfurred, same length or longer than head and body

DIET
Shoots, seeds, fruits and invertebrates

BREEDING
Age at first breeding: 8–12 weeks; breeding season: year-round; number of young: 3 to 12; gestation period: 25 days; breeding interval: 5 to 7 litters per year

LIFE SPAN
Up to 18 months

HABITAT
Grassland, forest, scrub, arid zones and agricultural habitats

DISTRIBUTION
North America south to northern Argentina, including Caribbean and many offshore island groups

STATUS
Many species declining; several species endangered or near-threatened

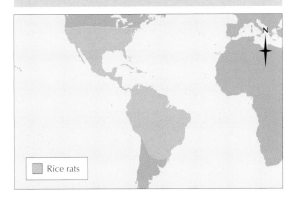

Rice rats

Rapid breeders

In the Tropics rice rats breed year-round. Females are sexually mature when they are about 7 weeks old and produce five in each litter. In the United States the species *O. palustris* produces fewer but larger litters of up to seven young. Gestation lasts about 25 days and the young are weaned at 11–13 days. In some species, mating may occur again only 10 hours after the birth of a litter.

Cause of rat plagues

The lemming is the most familiar example of a rodent that periodically becomes so abundant that it becomes a plague. In northern latitudes, plagues of rodents generally occur in cycles, the population suddenly building to astronomical proportions and then, equally suddenly, collapsing again. Plagues of Brazilian rice rats occur at long but regular intervals, and scientists have discovered that they correspond with the flowering of the taquara, the native bamboo, the largest of which grow to heights of 50–60 feet (15.2–18.3 m). When the taquaras flower, the ground becomes covered with their seeds, known as "taquara rice." Some types of taquaras seed every 11 years and others every 32 years, on average, but when they do, the rats feed on the fallen seeds and rapidly increase in numbers. When the seeds are finished, the rats move into fields and granaries seeking food.

Many species of rice rats are declining. Of the genus Oryzomys, *three are near-threatened, two are endangered and one has become extinct.*

Index

Page numbers in *italics* refer to picture captions.
Index entries in **bold** refer to guidepost or biome and habitat articles.

Page numbers in *italics* refer to picture captions. Index entries in **bold** refer to guidepost or biome and habitat articles.